SUPERCONDUCTING
INTERCALATED GRAPHITE

SUPERCONDUCTING INTERCALATED GRAPHITE

NICOLAS EMERY, CLAIRE HEROLD
AND
PHILIPPE LAGRANGE

Nova Science Publishers, Inc.
New York

For permission to use material from this book please contact us:
Telephone 631-231-7269; Fax 631-231-8175
Web Site: http://www.novapublishers.com

NOTICE TO THE READER

LIBRARY OF CONGRESS CATALOGING-IN-PUBLICATION DATA

ISBN: 978-1-60456-609-3

Published by Nova Science Publishers, Inc. ✦ New York

CONTENTS

PREFACE

The discovery in 2005 of superconductivity in YbC_6 and CaC_6, with substantially higher critical temperatures than the previously observed among the family of the graphite intercalation compounds, has largely renewed the interest for these well known lamellar compounds. Indeed, these critical temperatures reach 6.5 and 11.5 K respectively for ytterbium- and calcium-graphite phases. It was consequently interesting to collect all the informations concerning the superconductivity of these compounds from the discovery of this phenomenon observed in the heavy alkali metals graphite intercalation compounds in 1965, insisting particularly on the recent advances in this research field. After a general introduction that describes all the carbon materials, which are extremely various with dimensionalities varying from 3 to 0, leading to their large aptitude for the insertion/intercalation reactions, the authors widely developed the case of graphite: chemical bonds, crystal and electronic structures, anisotropy and ability to become a host structure. The authors insist on its strong anisotropy of chemical reactivity that allows the synthesis of very numerous intercalation compounds. The distinctive features of the intercalation reaction into graphite are reviewed (systematic charge transfer, staging, etc...) and are particularly developed in the case of the donor-type intercalation compounds, among which is precisely observed the superconductivity. For the latter, the various synthesis methods are successively described, showing the best route to use in order to obtain each type of compound. Then the authors review with detail the binary compounds, emphasizing their distinctive crystal and electronic structures and also their transport properties. The authors describe the superconductivity of all the compounds belonging to this family and showing this property. In the last part, the authors compare these superconducting binary intercalated graphite compounds with other lamellar superconductor: magnesium diboride. The ternary compounds are then studied, and the poly-layered nature of their intercalated sheets is given special attention. Their distinctive electronic structure is presented and their superconducting properties are described.

Chapter 1

GENERAL INTRODUCTION ABOUT CARBON MATERIALS

Carbon possesses a very particular position in the classification of the elements: indeed, it is the head of the central column of the periodic table. And this 14th column appears in fact as the spine of the classification. Its electronegativity is medium, and during the chemical reactions, it can be as well electron donor as electron acceptor, according to the cases.

On the other hand, it exhibits a very large ability to create some chains by bonding with itself. It is well known that the organic chemistry is born from this remarkable property. But it exists also some important consequences of this latter in the field of the inorganic chemistry. Indeed, the extreme variety of its allotropy is partly due to this property. It is due also to the ease for the carbon atom to change hybridization : sp^3, sp^2, sp, and even in some cases sp^x with x included between 2 and 3 (Figure 1).

The elemental solids obtained from sp^3 carbon atoms are diamond (cubic variety) or more rarely lonsdaleite (hexagonal variety). Both materials are three-dimensional, because they exhibit strong covalent bonds, that grow to infinity in the three directions of the space.

But, in the room conditions, the thermodynamically most stable carbon material is graphite, that appears as a lamellar solid. It is built from sp^2 carbon atoms and it is two-dimensional, because its covalent bonds grow to infinity in two directions of the space only : indeed the graphene planes are of course covalent structures, but they are stacked along the third axis by the means of very weak Van der Waals's bonds. One knows two graphite varieties : the hexagonal one, whose stacking is ABABAB..., and the rhomboedral one with an ABCABC... stacking. The first one is slightly more stable than the second one.

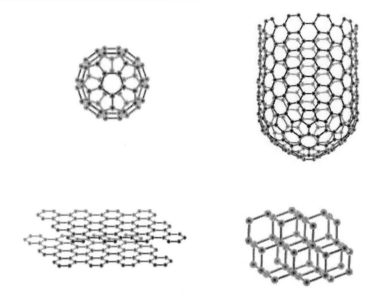

Figure 1. allotropy of carbon materials: fullerene, nanotube, graphite and diamond.

With sp hybridization, the carbon atoms lead to several one-dimensional solids called « chaoïtes ». They exhibit a fiber structure, since their covalent bonds grow to infinity in an unique direction of the space only. These linear covalent structures are gathered in beams.

All these carbon materials can be observed in the nature, but the chaoïtes are however particularly rare, due to their weaker thermodynamical stability.

Several other carbon materials derive from distorted graphene planes. Indeed, a perfect graphene plane is strictly flat, but if several hexagons are replaced by pentagons, it becomes convex and can turn into a closed structure. For this reason, it is admitted that the carbon hybridization in this case is included between 2 and 3. Thus, this phenomenon generates the class of fullerenes, among which the roughly spherical (truncated icosahedron) C_{60} molecule is the best known. Using Van der Waals's bonds, the assembly of numerous C_{60} molecules leads to a cubic solid, that appears as zero-dimensional, because its covalent bonds do not grow in any direction. This material is called fullerite and it is of course rather volatile, because it contains rather small molecules linked by Van der Waals's bonds.

Lastly, it is also possible for a graphene plane to wind around itself, after having suffered a more or less important torsion, leading to a cylindrical structure. These objects constitute the class of carbon nanotubes. They can be single-walled (the cylinder is unique) or multi-walled (several cylinders are fitted together), and they are associated within beams generated by Van der Waals's bonds. Of course, these carbon nanotubes, whose diameter is nanometric, appear as one-dimensional materials.

Save the 3D diamond structure, all these carbon materials are anisotropic and exhibit Van der Waals's bonds, that appears as weak points concerning the cohesion of these solids. Soft chemical reactions can exist for the latter, because, in these cases, the chemical reagents attack exclusively the areas of weak cohesion of the materials (often called Van der Waals's gaps), without disrupting their covalent parts. These gaps simply spread apart in order to create the space necessary for setting up the reagent. These specific soft reactions are called insertion or intercalation reactions, according to the cases : insertion preferentially corresponds to reactions without appreciable dilation for the pristine material, while intercalation makes use on the contrary of large dilation. In most cases, these insertion/intercalation reactions are reversible and it is possible to regenerate the pristine carbon material by heating.

Because of its medium electronegativity, we have seen that carbon appears as an amphoteric element. Indeed it is able to give or to pull out electrons, when it reacts with an other chemical species. This phenomenon turns up also in the case of the various allotropic carbon materials during these soft insertion/intercalation reactions. The donor or acceptor character is more or less pronounced according to the nature of the carbon material. However, whatever its range, the electron transfer is generally compulsory, otherwise the reaction doesn't take place at all.

With fullerites, the insertion word is more convenient, because the dilation remains often weak, due to the very large size of the tetrahedral and octahedral sites of the fcc structure of C_{60} fullerite, used for putting up the reagent. The intercalation word is more suitable for graphite, carbon nanotubes and chaoïtes, because the dilation can become considerable. It is precisely the case of graphite, that is the more studied among these carbon structures : its interplanar galleries spread apart very widely in order to accommodate sometimes very thick chemical species.

By direct experiments, we have shown, on the other hand, that, towards strong electron donors as alkali metals, C_{60} fullerite is more acceptor than graphite, that is to say C_{60} appears as the most oxidizing of both materials [1].

Chapter 2

GRAPHITE: VARIETY, BONDS, CRYSTALLINE AND ELECTRONIC STRUCTURES, ANISOTROPY AND MATERIAL CONSIDERED AS A HOST STRUCTURE

Graphite appears as the most stable carbon variety in room conditions. Its stability domain is especially extensive since the coordinates of the triple graphite-liquid-vapour point in its Clapeyron's diagram are close to 4100 K and 125 kbar [2].

The sp^2 hybridised carbon atoms that form the graphene planes are closely bound one another by means of very strong covalent bonds, whose length reaches 142 pm and energy 25 eV.mol^{-1} [3]. The strength of these bonds is revealed also by a very high sublimation point of about 3700 K. In this 2D structure, each carbon atom is associated with three coplanar neighbours, so that the value of the C-C-C angles is exactly 120°, according to the sp^2 hybridisation. The unused p_z orbital of each carbon can build with neighbouring atoms π_z molecular orbitals, that are of course delocalised on the whole of the graphene plane, as they are also in the case of the flat benzene molecule.

On the other hand, very weak Van der Waals's bonds provide the cohesion between the successive graphene planes, that are stacked up into graphite. For this reason, two successive graphene layers in graphite are 335 pm apart. But, they are not exactly superimposed and their stacking corresponds to the ABAB... sequence (Figure 2). The unit cell is hexagonal and belongs to the P6$_3$/mmc space group with the following parameters : a = 244 pm and c = 670 pm [4]. The carbon positions are : (000) and (2/3 1/3 0). And, taking the helicoïdal 6$_3$ axis into account, it appears four carbon atoms in each unit cell. In the end, graphite exhibits a very low density of 2.3.

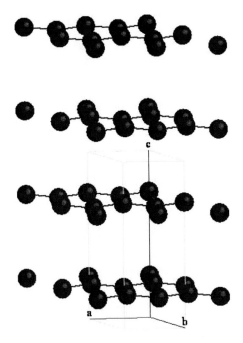

Figure 2. crystal structure of hexagonal graphite ($P6_3/mmc$).

This unusual structure confers a very strong 2D character on graphite. It is a perfect example of lamellar material, whose sheets are monolayered. Its anisotropy appears in all fields. Its mechanical properties for instance exhibit a very good aptitude for the cleavage, so that graphite appears as a material often used in lubricating (oils, pencil lead, etc…). Similarly, its electrical properties are very anisotropic : graphite is indeed a poor conductor along the **c**-axis and it is much better conductor in the other directions, since the corresponding resistivities reach respectively 0.1-1 Ω.cm and 40 $\mu\Omega$.cm [5].

Graphite can be either natural or synthetic. Well crystallised natural graphite platelets principally come from Madagascar, Sri Lanka, Russia or China. They are often mixed with other minerals like calcite or quartz and they have to be chemically purified after a manual sorting. Powder of synthetic graphite is obtained from pyrolysis of organic precursors followed by a step of graphitization. The ability of carbon to graphitize is determined during the pyrolysis. Hard carbon that is not graphitizable comes from carbonization without passing through a liquid phase. In this case, the number of graphitic slides is limited, with an distance larger than that of graphite (340-345 pm instead of 335 pm). Moreover these small crystallites are disoriented. Even after a heat treatment at 2500-3000°C, this carbon remains hard and without lubricant properties. On the contrary, when a liquid phase appears during the pyrolysis, crystallites have quite the same orientation and they are able to grow with a three dimensional organization. The graphitization is obtained by a heat treatment in the course of which most of defects disappear, the average diameter of crystallites increases and the distance between graphitic planes comes near 335 pm like in the perfect crystal. Artificial graphite powder with various granulometry can be obtained so.

Otherwise, it is also possible to prepare pyrolytic graphite. A carbon deposit on a heated graphite substrate is obtained by cracking of gaseous hydrocarbide diluted in argon. Graphitic sheets are parallel to the surface and when this pyrographite is heated at very high temperature (3000°C) under high pressure, the anisotropy of the material is increased and it becomes "highly oriented pyrographite" or "HOPG" whose properties are very close to those of a single crystal. In "HOPG", the c-axis of all crystallites, that are perpendicular to the graphene sheets, are parallel between them, with a maximal defect of 1°. However, the a and b axis are randomly oriented in the graphitic layers so that a HOPG platelet can be considered as a single crystal in the **c**-axis and as a powder in the perpendicular plane.

An other variety of graphite called "grafoil" is obtained by compression of first exfoliated graphite [2].

Chapter 3

CHEMICAL ANISOTROPY

Because of the nature of its bonds, the chemical properties of graphite are particularly anisotropic. In a first time, the presence of Van der Waals's bonds explains the chemical reactivity of graphite, by comparison with the diamond one, that is almost nonexistent : for instance, the oxidation of graphite is rather easy, whereas it is especially difficult in the case of diamond. In a second time, the soft reagents, that are unable to destroy the covalent bonds inside the graphene planes, take advantage of the weakness of the Van der Waals's bonds, in order to fill the interlayered galleries, leading to graphite intercalation compounds.

Because these reactions are necessarily oxido-reduction reactions, a charge transfer takes place systematically between the graphene planes and the reagent. Indeed, it is well established that the intercalation into graphite is strictly impossible without electron transfer. As graphite is amphoteric (now oxidizing, now reducing), it can be, according to the cases, electron acceptor or electron donor. But, it reacts only with strong enough reducing or oxidizing reagents.

Alkali metals, alkaline-earth metals and several lanthanides are very good reducing species, so that they intercalate easily into graphite at low temperature, releasing electrons into the graphene layers. This electron input creates a small dilation of the carbon-carbon distance in the covalent bonds. With these various metals, the intercalated sheets are strictly mono-atomic layers in all cases. The space between both graphene planes that surround the intercalated sheet widely varies according to the size of the intercalant. It is called « interplanar distance » (d_i).

On the other hand, if the reagent amount is not enough to fill up all the Van der Waals's galleries, only a few of them are full, while the others remain empty. It is impossible indeed for the intercalant to be diluted homogeneously in all Van der Waals's gaps. In these cases, the empty intervals and the full ones follow periodically one another along the c-axis, so that the full galleries attempt to scatter at the most. This specific phenomenon leads to define a « stage » (s) for each intercalation compound : it is the number of graphene layers included between two successive intercalated sheets (Figure 3.a). The stage one is thus characteristic of the saturated compounds, but it is often possible to synthesise some compounds belonging to the stages 2, 3, 4, 5, etc…, with potassium for instance. It is usual to call « c-axis repeat distance » (I_c) the space between two successive intercalated sheets. Thus the existence of high stage graphite intercalation compounds reveals some interactions between intercalated sheets through long distances.

The intercalated sheets can be less dense in the high stage compounds than in the first stage one. Indeed, KC_8 is the chemical formula of the first stage compound whereas KC_{12s} corresponds to the stage s compound. If we consider an intercalation reaction into graphite, the intercalant leads, in a first time, to the formation of high stages compounds, and then the progressive inlet of new amounts of reagent allows to obtain some compounds, whose stage regularly decreases as the intercalant takes up the empty galleries, according to the Daumas-Hérold model (Figure 3.b). This uncommon phenomenon is called « staging » and illustrates clearly the very high flexibility of the graphene plane. Indeed, among all the lamellar phases able to behave as host structure, the mono-layered graphene plane is truly the most flexible of them.

On the other hand, it is interesting to compare the density of the intercalated sheets contained in the various first stage intercalation compounds [6]. With the little atoms (Li, Ca, Sr, Ba, Yb, Eu…), the latter are especially dense, since their chemical formula corresponds to MC_6.

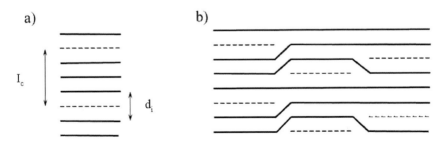

Figure 3. a) model of a graphite intercalation compound (stage 3) b) Daumas-Hérold model.

On the contrary, in the case of the biggest atoms (K, Rb, Cs), the formula becomes MC_8. For all these compounds, it is well established that the characteristic ABAB... graphitic sequence disappears for the benefit of the AAA... stacking, that leads to generate interlayered hexagonal prismatic sites, that the metallic atoms can very easily take up. All these data show clearly that the intercalated mono-layered sheets of these binary compounds are strictly commensurate with respect to the graphene planes. In all cases, the 2D unit cell of the intercalated sheets is consequently hexagonal. But, concerning the 3D one, it can be hexagonal, rhombohedral or even orthorhombic, according to the **c**-axis stacking mode of the successive intercalated sheets.

Associated with alkali metals, numerous other elements are able to intercalate into graphite, leading to ternary compounds [7]. We find in this category weakly electropositive elements as hydrogen, mercury, thallium, bismuth, etc... or strongly electronegative ones as oxygen, halogens, sulphur, etc... In all cases, these binary intercalated species behave towards graphite as electron donors and lead generally to poly-layered intercalated sheets. We will speak about these ternary phases at length in the subsequent developments.

On the other hand, very numerous oxidizing species are able to intercalate into graphite [8]. For instance, chlorine and iodine intercalate without difficulties. But various metallic halogenides (iron, copper, nickel, lanthanides chlorides...), several oxacids (H_2SO_4, HNO_3), some oxides (Cl_2O_7, SO_3, CrO_3...) can also intercalate easily into graphite.

All these chemical species intercalate while removing some electrons of graphene planes. This electronic removal causes a weak decrease in their carbon-carbon bond length. On the other hand, the previous « staging » phenomenon appears as well with these electron acceptors. It can be even particularly pronounced, since it has been reported a tenth stage graphite-sulphuric acid intercalation compound. Most of these compounds possess poly-layered intercalated sheets [9].

As examples, we find Cl-Fe-Cl and F-As-F three-layered sheets respectively in graphite-$FeCl_3$ and graphite-AsF_5 compounds, or mixed (Au-Cl) two-layered sheets in graphite-Au_2Cl_6 ones. In fact, such poly-layered sheets pre-exist before in the binary pristine halides, if the latter are solid in the room conditions. In these cases, they simply set up inside the Van der Waals's galleries, after the charge transfer occurs.

Consequently, these intercalated sheets are generally no-commensurate with respect to the adjacent graphene planes, that keep the ABAB... stacking of pristine graphite. In the end, it appears that both graphitic and intercalated sublattices grow parallely, using quite reduced interactions, so that both 2D unit cells remain very often quasi independent.

To sum up, in order to know the main features of a given graphite intercalation compound, it is absolutely necessary to specify its chemical formula, its stage, its repeat distance and, if possible, the c-axis atomic stacking inside its intercalated sheets and the corresponding 2D unit cell. But it is not easy, for very numerous compounds, to know their 3D unit cells.

Chapter 4

DONOR-TYPE GRAPHITE INTERCALATION COMPOUNDS

From this part, we will consider only the graphite-electron donors compounds, because the superconducting phases appear solely in this specific category.

The intercalation into graphite of the electropositive elements leads to a charge transfer, that releases electrons in the graphitic sublattice. The graphene plane appears as a macro-anion and the intercalated layer is made up of cations. For this reason, it is legitimate to regard these graphite intercalation compounds as « metallic graphitides ». But, of course, the charge transfer is not complete and the s valence electron remains partially tied to metallic atom. It is possible to say that the graphitide macro-anion constitutes the reducing part of the compound. On the whole, these lamellar compounds appear as strongly reducing species and cannot be handle to the air on pain of a severe oxidation.

When it is possible, the best synthesis method consists in putting the graphite material in the presence of the metal vapour, after having carefully evacuate air from the reaction tube. Thus, graphite and metal are arranged in each end of the glass tube sealed under vacuum and a temperature gradient is carried out between both reagents (Figure 4) [10]. The lower temperature is assigned to the metal, so that it imposes its vapour pressure on the whole of the tube. According to the value of the higher temperature, it is possible to synthesise first, second, third... stage binary compounds. The stage increases when simultaneously increases the temperature gradient range [6].

This method allows to prepare very pure graphite intercalation compounds, but, the vapour pressure of the metal has to be sufficiently high, even at low temperatures (200-300°C).

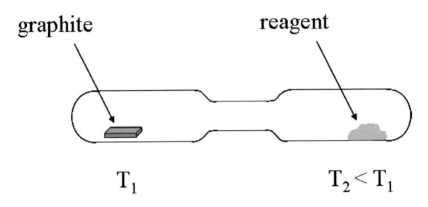

Figure 4. two-bulb tube for vapour phase intercalation reaction.

It is the case for the heavy alkali metals (K, Rb and Cs), but all the other metallic elements able to intercalate into graphite need a too high temperature, so that the reaction becoming too violent causes the destruction of the graphene planes and leads to the formation of metallic carbides. Consequently, for these metals (Li, Ca, Ba, Eu, Yb...), other synthesis methods have to be carried out. Two well established routes are recommended : solid-solid and liquid-solid reactions.

The solid-solid reactions require to use a powder mixture of pure metal and pristine graphite. They have been particularly used in order to synthesise the various graphite-lithium phases [11]. Both well mixed metal and graphite powders are compressed in a pure argon atmosphere until 10 kPa and then heated under Argon or vacuum at 200°C during 24 hours. With lithium, the reaction product is LiC_x, according to the initial Li : C ratio. Of course, the reaction temperature allows the easy lithium intercalation but remains quite low in order to avoid the formation of lithium acetylide (Li_2C_2).

On the other hand, it is well established that temperature and pressure induced by the shocks occurring during a ball-milling can temporarily reach very high values (close to the previous conditions). Therefore, the grinding of graphite and lithium powders mixture can be used in order to synthesize graphite-lithium compounds [12]. Small amounts of n-dodecane ($C_{12}H_{26}$), which is entirely inert towards lithium, are added in order to avoid agglomerating lithium particles on the milling tools during the grinding reaction.

The liquid-solid reactions are mostly more complex and especially more delicate, because it is necessary to liquefy the metallic reagent. If the melting point of the metal is low, it is easy of course to obtain a liquid reagent, but, when it is high, we have to use a mixture with a second metal in order to decrease the melting temperature. Indeed, the intercalation reaction temperature has to remain low in order to avoid the destruction of the graphene planes. The choice of the added metal is especially difficult, because, when the alloy reacts with graphite, only the first metal has to intercalate. Now, for a given alloy, reaction temperature and alloy composition strongly influence the nature of the reaction product. It is often very difficult to find the best reaction conditions able to lead to a binary intercalation compound. Sometimes, they do not exist and it is thus only possible to obtain a ternary intercalation compound or even no intercalation at all.

Consequently controlling three basic parameters appears as very important in order to succeed in the intercalation into graphite of a metallic element by reaction with a liquid alloy : it is a matter of the nature of the associated metal, its concentration and the reaction temperature. For instance, LiC_6 can be prepared easily using a Na-Li liquid alloy in well chosen conditions [13], CaC_6 is obtained from a Ca-Li liquid alloy [14]. Of course, these reactions are carried out safe from air (under vacuum or argon atmosphere, according to the cases). After reaction, separating the solid compound from the liquid alloy is easy only if pristine graphite is a pyrolytic graphite platelet. For this reason, it is not recommended to carry out such a reaction using graphite powder.

Concerning the ternary graphite-electron donors compounds, they contain two intercalated elements, whose first of them has to be metallic, while the second one can be metallic or not according to the cases. As previously, the charge transfer necessarily occurs with an increase of the negative charges inside the graphene planes, which are reduced.

When is intercalated a mixture of two strongly electropositive metals (K-Rb, Rb-Cs, K-Cs) [15-17], the intercalant remains mono-layered, as in the binary compounds. In all other cases, on the contrary, the ternary compounds contain poly-layered intercalated sheets. More frequently, three atomic layers are intercalated between the graphene planes and this sandwich possesses a central layer of less electropositive atoms surrounded by two symmetrical layers, that are made of strongly electropositive atoms [18-19].

In order to synthesise these ternaries, the usual method is the liquid-solid reaction. A pyrolytic graphite platelet is plunged in a liquid metallic alloy or in a melted metal containing very small amounts of the second element, when this latter is weakly electropositive or strongly electronegative.

In exceptional cases, it is easier to carry out the synthesis by means of two successive steps [20] (Figure 5). In a first time, is prepared a first stage potassium-graphite, for instance, by the classical vapour-solid method, and, in a second time, this binary phase is put in the presence of hydrogen or mercury vapour.

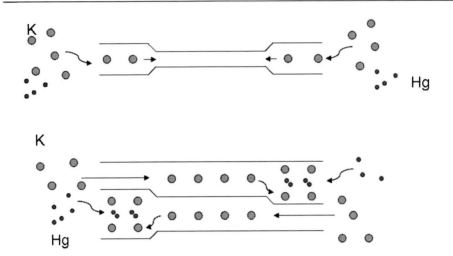

Figure 5. intercalation by means of two successive steps (case of potassium amalgams).

This second intercalation reaction causes an increase in the stage (from 1 to 2), and simultaneously the mono-layered intercalated sheets become three-layers. Only intercalation compounds of higher than one stage can be obtained using this method.

Chapter 5

BINARY GRAPHITE INTERCALATION COMPOUNDS

Alkali metals graphite intercalation compounds are particularly well known, because they were the most studied among the lamellar graphite compounds.

Potassium, rubidium and cesium (known as heavy alkali metals) intercalate very easily into graphite galleries [21]. They lead to intercalation phases, whose stage can vary from 1 to 2, 3, etc... The first stage compounds exhibit a specific brilliant gold colour. Their chemical formulas are KC_8, RbC_8 and CsC_8, and their interplanar distances reach respectively 535, 565 and 592 pm, according to the increasing size of the alkali metals. Their crystal structures exhibit very pronounced likenesses [22-24]. Of course, their intercalated sheets possess exactly the same geometry : each intercalated metallic atom occupies a prismatic hexagonal site, due to the AAA... c-axis stacking of the graphene planes. But, in each gallery, the quarter of the hexagonal sites only are occupied, so that it is possible to define four distinct sites into the Van der Waals's gap : α, β, γ and δ. Although the graphene planes are exactly superimposed, the metal layers, on the contrary, avoid strictly superimposing at the most, due to the Coulomb's repulsion. For potassium and rubidium, the c-axis sequence is AαAβAγAδ..., revealing long distance interactions between intercalated sheets. Consequently, the c parameters of KC_8 and RbC_8 are respectively 2140 and 2260 pm. Because of its higher interplanar distance, CsC_8 adopts a shorter sequence AαAβAγ..., so that its c parameter reaches only 1776 pm. These differencies lead crystal structures, whose geometries are orthorhombic [2, 3] for KC_8 and RbC_8 (with Fddd space group, Figure 6) and hexagonal [4] for CsC_8 (with $P6_422$ space group, Figure 7).

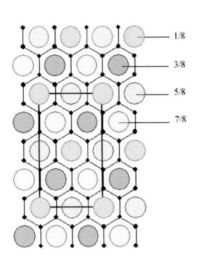

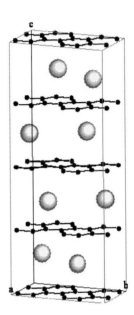

Figure 6. crystal structure of KC_8 (Fddd).

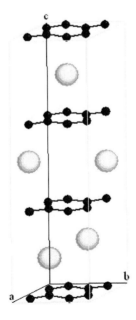

Figure 7. crystal structure of CsC_8 (P6$_4$22).

It is harder to synthesise sodium graphite intercalation compounds. Indeed, only high stage compounds [25] are reported, whose formulas are NaC_{8s}, with stage number s = 4 to 8. The reason for the difficulty in the synthesis of these compounds consists in two competing factors : the ionisation potential of sodium and its size. The ionisation potential increases from Cs to Li (from 3.9 eV to 5.4 eV), whereas the ionic radius strongly decreases (from 200 pm for Cs to 100 pm for Li). Of course, the lower the ionisation potential, the easier is the electron transfer from metal towards graphite. Consequently, the energy gain upon the charge transfer becomes more and more weak from Cs to Li. That is the reason that explains why a first stage sodium graphite compound is with difficulty thermodynamically stabilized.

According to this argument, it can be surprising that Li is able to give without difficulty a first stage LiC_6 graphite intercalation compound [11]. This observation can be explained by the very small size of Li^+. Indeed, owing to its size, lithium causes a weak spacing of the graphitic galleries, corresponding to a small energy, that promotes the stability of the intercalation compounds, in spite of the previous unfavourable factor. In the case of sodium, that is an intermediate element, both factors are simultaneously unfavourable : weak ionisation potential and too large spacing of the carbon layers.

In LiC_6, the interplanar distance is very small (370 pm), because of the small size of Li^+. The lithium atoms occupy prismatic hexagonal sites, as previously in the case of heavy alkali metals compounds. But with lithium, the intercalated sheets are more dense, so that, in each of them the third of the hexagonal sites contains a lithium atom. Consequently, we have to define only three distinct sites into the graphitic galleries : α, β and γ. In spite of that, the metallic atoms are exactly superimposed along the **c**-axis, according to the AαAα... sequence (the repeat distance is of course equal to the interplanar distance). Such a stacking makes two Li atoms very close through the graphene plane (370 pm), so that one can consider that exactly **c**-axis superimposed Li atoms are partially themselves bound by means of covalent bonds. Instead of a Coulomb's repulsion as previously, it is, on the contrary, a covalent attraction that governs the crystal structure. The unit cell of LiC_6 is hexagonal and its space group is P6/mmm [11] (Figure 8).

Alkaline earth metals intercalate into graphite similarly to lithium. Indeed, calcium, strontium and barium [26] lead to first stage phases, whose formulas are respectively CaC_6, SrC_6 and BaC_6. The interplanar distances increase regularly from Ca to Ba, as the corresponding ionic radii : 455, 494 and 525 pm. These metals exhibit high ionization potentials and simultaneously low vapour pressures, so that the intercalation by solid-vapour reactions appear as rather difficult. Thus, it is recommended to carry out the intercalation of these metals by solid-liquid reactions, using Li-Ca and Li-Ba well chosen alloys [27, 28].

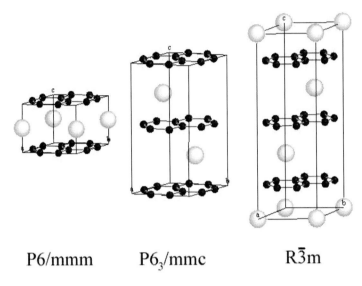

$$P6/mmm \qquad P6_3/mmc \qquad R\bar{3}m$$

Figure 8. crystal structures of MC_6 compounds.

The 2D structure of the intercalated sheets is exactly the same as for LiC_6. But, due to larger ionic radii, the Coulomb's repulsions lead to **c**-axis stackings quite different : AαAβ... and AαAβAγ... respectively for BaC_6 and SrC_6 [26] on the one hand and for CaC_6 [29] on the other hand. The first stacking leads to an hexagonal unit cell (space group : $P6_3/mmc$) and the second one to a rhomboedric cell (space group : R-3m) (Figure 8).

Several of the rare earth metals (Eu, Yb, Sm) have been reported to be intercalated into graphite [30]. As previously, high ionization potentials and low vapour pressures lead to difficult solid-vapour intercalation reactions. When it is possible, the solid-liquid one is preferable. Thus, in the case of europium, well chosen Li-Eu alloys are able to give first stage graphite-Eu intercalation compound. The formula of these lamellar phases is MC_6, due to rather small ionic radii. And their interplanar distances reach 457, 487 and 471 pm respectively for YbC_6, EuC_6 and SmC_6. The crystal structure of EuC_6 and YbC_6 corresponds to the **c**-axis AαAβ... sequence, so that their unit cell is hexagonal (space group : $P6_3/mmc$).

High-pressure synthesis allows to prepare highly dense alkali metal graphite intercalation compounds. In these lamellar phases, the alkali metal densities can exceed considerably the maximum value observed for graphite compounds obtained by more traditional methods.

Lithium gives, beyond 8 kbar, very high density compound LiC_2 [31, 32]. But this latter is decomposed into less dense lamellar phases as the pressure is released. Thus, it is possible to identify successively the following compounds : $Li_{11}C_{24}$, Li_9C_{24} and Li_7C_{24}. In spite of its rather high density, Li_7C_{24} remains stable even at ambient pressure. On the other hand, using the ball-milling method, the LiC_3 metastable phase was obtained [12].

Sodium, which is not able to give first stage intercalation compound using classical methods, leads under high pressure (35-40 kbar) to dense NaC_{2-3} lamellar phases [33]. But they decompose in graphite and metal below 20 kbar.

In the case of heavy alkali metals, KC_4, $RbC_{4.5}$ and CsC_4 were reported [34, 35]. These various intercalation compounds are obtained respectively beyond 5 kbar, at 20-25 kbar and at 2 kbar. Among them, only CsC_4 remains metastable as the pressure is released, although it is far from equilibrium conditions.

The intercalated sheets for highly dense lithium, potassium, rubidium and cesium graphite intercalation compounds are mono-layers, with classical previously reported interplanar distances. On the contrary, it is not the case with sodium, since $NaC_{2.6}$ consists of three-layered metal intercalated sheets and exhibits an interplanar distance of 704 pm [33]. No information concerning its in-plane structure was reported.

In LiC_2, the lithium atoms occupy all the hexagonal prismatic sites derived from the AAA... c-axis stacking of the graphene planes, so that the in-plane Li-Li interatomic distance is 248 pm. This value is especially weak, since even in cubic centered pristine lithium metal, the Li-Li distance is larger (309 pm). The Li_7C_{24} compound following the release of pressure exhibits hexagonal 2D Li_7 clusters [32] (Figure 9), leading to an in-plane hexagonal unit cell, whose a parameter of 863 pm is equal to $2\sqrt{3}.a_G$ (a_G is the hexagonal in-plane parameter of graphene plane). Along the c-axis, its c parameter is three times higher than its repeat distance (1110 pm). The presence of lithium clusters, which are considered to be stabilized with the formation of Li-Li covalent bonds, appear of course as the consequence of the application of pressure.

The crystal structure of the ball-milled LiC_3 compound exhibits an unusual small splitting of the intercalated lithium layer [36]. Indeed, in its hexagonal unit cell, that belongs to the $P6_3/mmc$ space group (Figure 10), with a = 430 pm and c = 740 pm (twice the repeat distance), it appears a shift of the intercalated lithium layer in two planes at ± 44 pm from the medium plane of the graphitic galleries.

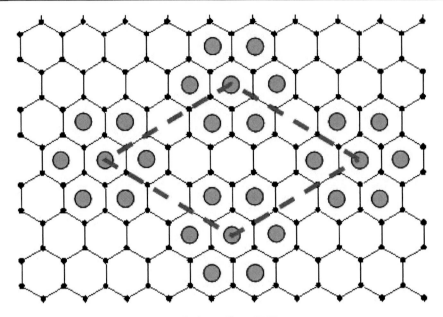

Figure 9. two-dimensional structure of Li_7C_{24} (from [32]).

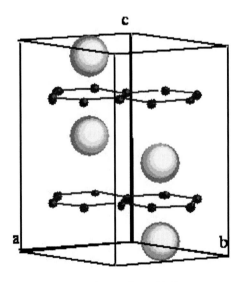

Figure 10. crystal structure of LiC_3 ($P6_3$/mmc, from [36]).

The metastable CsC_4 structure has been reported [37]. The in-plane unit cell of its intercalated sheets is rectangular with $a = 2\sqrt{3}.a_G$ and $b = 2a_G$ (Figure 11).

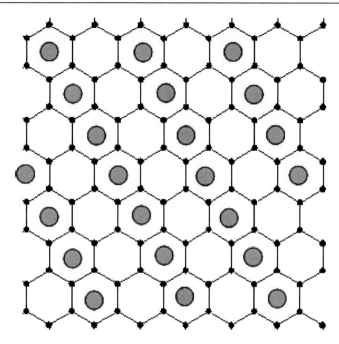

Figure 11. two-dimensional structure of CsC_4 (from [37]).

In this compound, cesium linear chains are formed by occupying half the hexagonal prismatic sites. In every chain, the Cs-Cs interatomic distance reaches 248 pm and two adjacent chains are one hexagon ring apart from each other, so that the interchain distance is 428 pm.

In pristine graphite material, the combination of sp^2 σ- and π-bonds is the origin of the very strong intralayer interactions, while the overlap of π-bonds between the successive graphene layers generates the weak interlayer interactions [5]. In examining the electronic properties around the Fermi energy, which is especially important for the electronic structure of the graphite intercalation compounds, the π-electron orbitals play an essential role. They provide graphite with its exceptional properties. Conversely, the σ-bands, which exhibit larger energy than π-bands, are located far from the Fermi energy. Consequently, they do not contribute to any noteworthy change in the electronic properties when intercalated sheets are introduced between the graphene layers.

On the basis of the tight binding model [38] and using Slonczewski-Weiss-McClure parameters, it was established that the electronic structure of pristine graphite shows four π-bands due to the ABAB... graphitic stacking : E_1, E_2 and E_3 (twofold degenerate). And the Fermi level crosses the E_3 band, producing one electron pocket and two hole one, so that graphite appears as a semi-metal. In spite of its relative simplicity, in most cases, this model is fully efficient in order to explain the electronic properties of graphite. The numbers of charge carriers (holes and electrons) are estimated at 3.10^{18} cm^{-3}. They are of course considerably smaller than for true metals. In addition, holes and electrons possess noteworthy smaller effective masses : 0.039m and 0.057m, respectively, where m is the free electron mass [39].

The simplest model of the electronic structure of graphite intercalation compounds is obtained from the π-band structure on the basis of the Slonczewski-Weiss-McClure tight binding model with the rigid bands scheme, where the interactions between the graphene sheets and the intercalated ones are neglected (Figure 12). It is known as Blinowski-Rigaux model [40, 41]. Because of its simplicity, it is only convenient for the electron acceptor graphite intercalation compounds. Indeed, the electronic states of the intercalated acceptor species are well below the Fermi level. The electrons in the valence π-band are transferred to the acceptor band, so that numerous holes are produced around the top of the π-band, able to play the role of charge carriers.

In the case of electron donor graphite intercalation compounds, the intercalate band (for example, the 6s level in the case of cesium) is located close to the Fermi level. Consequently, this intercalate band has to be taken into account in addition to the graphitic π*-band in the working out of the corresponding electronic structure. Of course, the latter is rather complicated compared to that of acceptor graphite intercalation compounds. In order to obtain convenient informations on the electronic structure of electron donor graphite intercalation compounds, it is thus necessary to use more sophisticated models than the Blinowski-Rigaux one.

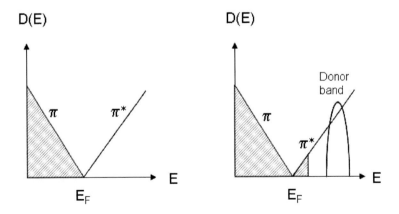

Figure 12. electronic structures of graphite and donor-type GIC (rigid bands).

LiC_6 has been a target of most intensive studies using numerous methods of calculation, because its electronic structure is the simplest among the electron donor graphite intercalation compounds [42-47]. The interlayer state of graphite, which is unoccupied in pristine graphite, plays an important role through hybridisation between interlayer state and intercalate one [46, 48-52]. An overlap between the interlayer band and the intercalate one occurs because both of them are located in the same energy range. The resulting hybridisation gives interlayer-intercalate bands featuring a combination of both bands in the graphitic interlayer space. Consequently, the intercalate electronic state is not purely that of intercalate, but possesses a strong contribution from the interlayer graphitic state. This interlayer-intercalate band exhibits a 3D nature, which is a noteworthy feature of the electronic structure of the electron donor graphite intercalation compounds (especially first stage ones). It is located between around 1.3 and 6.3 eV, that is to say well above the Fermi level, so that the electronic structure of LiC_6 can be reasonably understood in terms of almost complete charge transfer between lithium atoms and graphene layers, leading to the presence of completely ionized Li^+ cations between the graphene planes. This result is absolutely consistent with that would predict the simple rigid band scheme, if we consider that the Li 2s level is far above the Fermi energy.

The behaviour of the first stage KC_8, RbC_8 and CsC_8 compounds is quite different, because it appears a close proximity of the interlayer-intercalate state and the graphitic π-band, so that the situation is rather complicated. Thus, it is useful to consider, in the case of KC_8 for example, the hybridization of the graphitic π-band and the K 4s band [53-54]. The calculations show that the graphitic π*-band is mixed with the K 4s one in the vicinity of the Fermi level, leading to only partially ionized potassium atoms ($K^{0.6+}$) in the graphitic galleries. The same phenomenon occurs in both RbC_8 and CsC_8 cases, the interlayer-intercalate band moving however to higher energy from potassium to cesium [52, 53, 55].

The alkaline earth and rare earth metals lead to first stage graphite intercalation compounds, whose formula is MC_6. Among these phases, BaC_6 was the most investigated [56-58]. Three bands cross its Fermi level : two of them are ascribed to the graphitic π-bands, while the third one is due to the interlayer-intercalate band. The partially occupied latter plays an important role in characterizing the electronic structure in the vicinity of the Fermi level. It contains also contributions of Ba 6s and 5d orbitals. Indeed, the hybridization between Ba 5d and graphitic π-bands has an important effect on the electronic structure of

BaC_6. It seems that one of both electrons in the Ba 6s band is transferred to graphitic π-band, while the second stays in the interlayer-intercalate one.

In EuC_6, that possesses a particular interest with regard to magnetism, occurs an interaction between Eu 4f and graphitic π-electrons, that plays an essential role, giving rise to novel magnetic behaviour in this compound [59-63]. The 4f electrons give localized states, which are responsible for localized magnetic moments, whereas the delocalized 6s electrons cause the formation of bondings with the graphitic π-electrons upon intercalation. It occurs an hybridization of 6s, 6p and 5d states of europium and appears simultaneously an overlap between the graphitic π-band and Eu spd orbitals. Consequently, a sd-hybrid band emerges as a partially filled state between the Fermi level and the π*-band. At last, the amount of charge transferred from Eu to graphite is estimated at 0.5. This small value is ascribed to covalent admixtures present in EuC_6 as well as the rather large electronegativity of europium (compared with alkali metals). Similar electronic features were observed in the case of YbC_6 compound.

Obtained by means of high-pressure synthesis, superdense alkali metal graphite intercalation compounds exhibit rather different features from classical binary phases. The application of pressure, leading to reduce the volume of the atoms, enhances simultaneously the contribution of p- or d-levels in the electronic structures, in spite of the predominance of the s-level [64, 65]. The p- and d-orbitals, which have directional features instead of a spherical symmetry as the s-ones, promote the covalent feature of the bonds. Thus, the presence of Li clusters (Li_7C_{24}) and Cs chains (CsC_4) appear as the consequence of the covalent bond formation in the intercalated layers of these compounds [37, 66]. If we consider the very large concentration of Li in the superdense lithium compounds, it appears nevertheless that the electron transfer rate per carbon atom from Li to C exhibits approximately the same value in these phases as in LiC_6. Consequently, the charge transfer per lithium atom is strongly reduced. This phenomenon can be explained, in Li_7C_{24} for instance, by an obvious participation of covalent bonds, having directional p-character, in the formation of the 2D Li_7 clusters. In CsC_4, it was also reported that the charge transfer per cesium atom is weaker than in CsC_8. This suggests the presence of Cs-Cs covalency, which is responsible for the formation of cesium chains, brought about by the presence of Cs 5d-orbitals [67].

It is well established that pristine host graphite exhibits a semi-metallic behaviour. In graphite intercalation compounds, the intercalated electron donors release a large concentration of charge carriers (electrons) in graphene layers, so that these compounds become truly metallic. Indeed, their characteristic in-plane conductivity reaches about 10^5 $\Omega^{-1}.cm^{-1}$ [68]. The 2D features of their electronic properties lead however to extremely different conductivity processes between in-plane and interplane directions.

Since these compounds appear as 2D materials, their electron transport properties are particularly modified from those observed for classical 3D metals. For instance, their Debye temperature, close to 2500 K, is strongly higher than that measured for ordinary metals (it reaches indeed in most cases only 300 K). The in-plane resistivity is expressed by the means of an empirical formula, whose temperature dependence is quadratic, even though it is simply linear in the case of ordinary 3D metals [69-72].

On the other hand, the **c**-axis conductivity is very weak in comparison with the in-plane one. A 2D band model, widely used in the case of electron acceptors graphite intercalation compounds, is not efficient for the electron donors ones because of too strong correlations between the graphene planes. Indeed, the intercalated sheets are usually electrically conductive in these cases, so that three dimensionality has to be taken into consideration in order to rightly analyse the **c**-axis conductivity of these phases. Indeed, it was established that donor intercalates contribute effectively to c-axis electron transport. As a consequence of this phenomenon, it appears that the c-axis conductivity of the binary electron donors graphite intercalation compounds is always larger than that of pristine graphite. Thus, HOPG, LiC_6 and KC_8 possess c-axis conductivities (σ_c in $\Omega^{-1}.cm^{-1}$) of 8.3, 1.8 x 10^4 and 1.94 x 10^3 respectively [73-75]. This is considered to be associated with the presence of the interlayer-intercalate states in the graphitic galleries, as it has been previously reported.

The partial three-dimensionality of the binary electron donors graphite intercalation compounds is also easily observed examining the room temperature σ_a/σ_c ratio. In the case of pristine graphite (HOPG), it reaches indeed 3 x 10^3, and frequently 10^4, 10^5 and even 10^6 for the electron acceptors compounds, but it falls to 56 and 14 for KC_8 and LiC_6 respectively.

Impurity- and phonon-assisted hopping seems to play a basic role in the behaviour of the c-axis conductivity, according to Sugihara [76-78]. On the other hand, Shimamura [79] claims the importance of the presence of conduction paths, which are produced by intercalation of incomplete layered sheets (Daumas-Hérold domains for instance).

Chapter 6

SUPERCONDUCTIVITY OF BINARY GRAPHITE INTERCALATION COMPOUNDS

Among graphite intercalation compounds (GICs), only donor type compounds exhibit superconducting properties.

The superconductivity in graphite intercalation compounds has been discovered in 1965 by Hannay et al. [80] in first stage alkali metal compounds. The transition temperatures observed range from 0.39 K to 0.55 K for KC_8, from 0.023 K to 0.151 K for RbC_8 and from 0.020 K to 0.135 K for CsC_8. The authors didn't observe any superconductivity down to 0.011 K for second stage compounds.

After this work, superconductivity of KC_8 was confirmed quite twenty years later by Koike et al. [81-83], by Kobayashi and Tsujikawa [84, 85], and the superconducting properties were investigated in detail. Later, the superconductivity of RbC_8 was confirmed by Kobayashi et al. [85, 86].

In the end of the eighties, the application of high pressure on mixtures of graphite and alkali metals have led to superdense GIC's, a new family of superconducting materials with a highest T_c of 6 K for CsC_4. But these compounds are not stable in ambient conditions (temperature and pressure). Twenty-five years later, the discovery of the superconductivity in YbC_6 and CaC_6 [87, 88] with critical temperature of 6.5 K and 11.5 K respectively has led to a renewed interest in graphite intercalation compounds.

In all systems, when the graphite intercalation compound is superconducting, the intercalated element doesn't superconduct.

All superconducting binary GICs are listed in Table I.

Table I. Superconducting binary first GICs

GIC	d_i (pm)	$H_{c2}^{\perp}/H_{c2}^{//}$	T_c (K)	References
KC$_8$	535	4.7 - 6.2	0.39 – 0.55 0.128 – 0.198 0.15	80 81-83 84, 85
RbC$_8$	565		0.023 – 0.151 0.026	80 86
CsC$_8$	592		0.020 – 0.135	80
LiC$_2$	368		1.9 0.9 - 1.6	94, 95 95
NaC$_2$			5	93, 95, 96
NaC$_3$	398.5 and 557.8 (at T =100 K)		2.3-3.8 2.3-4.5	93, 96 95
NaC$_4$			2-3.5	95
KC$_3$			3	93, 95
KC$_4$			5.5 0.35-3	97 95
KC$_6$	530.5 (at T =100 K)		1.5 1.5 1.45-1.55	93, 97 95 34
RbC$_{4.5}$			1.6	35
CsC$_4$			6	98
YbC$_6$	457	2	6.5	87
CaC$_6$	451 452.4	2 3.5-4	11.5 11.5 11.3	87 88 105
SrC$_6$	495	2	1.65	124

KC$_8$ COMPOUND

Most studies were performed on potassium-graphite compounds. Superconducting properties of this system were investigated by Koike et al. by low frequency a.c. magnetic susceptibility and electrical resistivity measurements using HOPG based samples. The intercalation was carried out using potassium in vapour phase. The superconducting transition that is quite narrow, appeared between 0.128 K and 0.198 K measured on 13 samples, leading to an average value close to 0.15 K. Magnetic susceptibility measurements performed with the applied magnetic field parallel or perpendicular to the c-axis revealed a remarkable anisotropy. When the applied field is parallel to the c-axis, KC$_8$

appeared as a type I superconductor whereas when the field is applied in the perpendicular plane, type II superconductivity is observed. An anisotropic behaviour is also evidenced by the values of the critical fields : H_{c2} for an applied field perpendicular to the c-axis is more than twice higher than H_c for an applied field parallel to the c-axis. This anisotropic feature declines with the decrease of the K/C ratio. However, the reduction of potassium concentration doesn't change the value of the transition temperature up to the composition $KC_{14.7}$. No transition was found down to 0.060 K for $KC_{16.7}$ and $KC_{21.6}$ [85] so that it is the confirmation that the second stage potassium-graphite compound is not superconducting. However, the transition temperature depends on the carbon materials used for the preparation of the samples. Indeed, T_c is 0.080 K when powdered pyrolytic graphite is used, 0.125 K using grafoil and in the best case, 0.162 K with HOPG [85]. The higher is the crystal structure of the host material, the higher is the T_c.

RbC_8 COMPOUND

In comparison, RbC_8 became superconductor at 0.026 K [86] and it appeared as a type I. As in the case of graphite-potassium system, in the graphite-rubidium one, superconductivity was evidenced only in the first stage compound.

HYDROGENATION OF KC_8

The hydrogenation of KC_8 has led to superconducting quasi-binary compounds whose sheets are monolayered. This property depends on hydrogen concentration, stage and host graphite. $KH_{0.19}C_8$ prepared by hydrogen chemisorption from HOPG based KC_8 is superconducting below 0.195 K [89]. This compound seems to be a mixture of $KH_{0.1}C_8$ (stage 1) and a stage 2 potassium hydride ternary compound that is not superconducting. The latter is described in VII. The superconductivity is then induced by the quasi-binary $KH_{0.1}C_8$. As in KC_8, $KH_{0.1}C_8$ appeared as a type I superconductor when the applied field is parallel to the c-axis whereas type II superconductivity is observed when the field is applied in the perpendicular plane. No superconductivity is observed in the ternary stage 2 compound $KH_{0.67}C_8$ prepared from grafoil.

KC_8 UNDER PRESSURE

The application of high pressure is often tried to increase the values of the transition temperature in superconducting materials.

In the case of KC_8, DeLong et al. [90, 91] found at an applied pressure ranging from 2 kbar to 15 kbar a first-order phase transition from a superconducting state with a T_c of 0.13 K to a an other one with a T_c of 1.7 K. The exact nature of the high pressure phase remains uncertain even if structural changes are the most probable.

Later Belash et al. [92, 93], succeeded in increasing the critical temperature of KC_8 applying pressure up to 13 kbar to 1.5 K. But with an applied pressure higher than 13 kbar, T_c decreases to 1.4 K at 30 kbar and to 1.13 K at 37 kbar.

DENSE ALKALI METAL COMPOUNDS

Graphite intercalation compounds with all alkali metals were obtained using a high pressure synthesis method. These high concentration alkali metals compounds form a group of superconducting materials with transition temperatures higher than those of classical binaries.

Most of these compounds are not stable under standard conditions so that they were quenched to temperatures of 90 K to 100 K and then stored in liquid nitrogen.

In the case of lithium, weighed amounts of graphite and lithium in order to obtain LiC_6, LiC_3 and LiC_2 were treated at 77 K under a pressure ranging from 30 to 40 kbar for 6 hours [93-95]. For LiC_2 samples, a superconducting transition was measured at 1.9 K, whereas no transition has been observed at a temperature higher than 0.35 K for LiC_6 and LiC_3.

The compositions NaC_4, NaC_3 and NaC_2 were studied [93, 95, 96]. All became superconducting after the application of the pressure. NaC_3 was synthesized at 227 K under 45 kbar for 0.5 h in a copper ampoule and was then placed in a teflon container and treated at room temperature under 35 kbar for 22 h. After quenching the sample in liquid nitrogen, X-ray diffraction was carried out at 100 K and two series of *001* reflexions were observed and attributed to the presence of two phases with repeat distances of 1115.6 ± 0.7 pm and 797.0 ± 1.5 pm.

As in the case of lithium, there is a dependence of the critical temperature to the alkali metal concentration. However, in both cases, T_c decreases monotonously with time, even when the samples are stored in liquid nitrogen.

The potassium GICs were the most studied compounds. In this system, the richest potassium compound is KC_3. It was obtained at 77 K under a pressure of 30-40 kbar for 6 hours. Less dense phases, KC_6 and KC_8 were prepared at 77 K with an applied pressure of 3-7 kbar for 6 hours. KC_8 samples are stable under normal conditions. The magnetic susceptibility of each potassium GIC versus temperature is shown in figure 6.1. The transition is abrupt with complete saturation for KC_3, KC_6 and KC_8, whereas that of KC_4 is probably due to a mixture of several superconducting phases (Figure13).

In these compounds, the T_c value is controlled by the potassium content since it increases when the C/K ratio decreases but it depends also on the metal ordering in the layer. Temperature and angular dependences of H_{c2} and H_{c1} for both KC_3 and KC_6 compounds, as well as H_{c2} anisotropy for KC_8 were studied [95]. All compounds can be described using the phenomenological anisotropic effective mass model suggesting that the increase in the potassium concentration makes the electronic structure more three dimensional.

For lithium, sodium and potassium dense compounds, the transition temperature raises when the alkali metal concentration increases. This suggests an important role of alkali metal electrons in enhancing superconductivity.

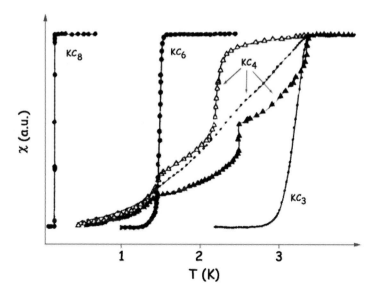

Figure 13. Magnetic susceptibility of KC_x GICs (from [95]).

Graphite intercalation compounds with the heaviest alkali metals, rubidium and cesium, were less studied than Li, Na, K compounds [35, 97].

Among all these alkali metals dense compounds, CsC_4 exhibits the highest critical temperature of 6 K [98].

RECENT ADVANCES IN CaC_6, SrC_6 AND YbC_6

In 2005, Weller et al. [87] reported a superconducting transition in YbC_6 and CaC_6 with respective T_c of 6.5 K and 11.5 K. In the case of CaC_6, in the magnetization results, no saturation of diamagnetism is obtained down to 2 K, due to the reduced samples quality. Emery et al. obtained the saturation of the diamagnetic signal after a very sharp transition [88] using bulk samples, prepared by a liquid-solid synthesis route [14]. Among the GICs, these compounds exhibit the highest transition temperatures so that a renewed interest in these materials appeared recently. Many efforts were done in experimental and theoretical studies in this field.

Two kinds of pairing mechanisms have been developed for understanding the origin of superconductivity in these materials [99-103]. A conventional electron-phonon interaction could be sufficient for Calandra and Mauri [99, 100] and for Mazin et al. [101, 102]. On the other hand, Csányi et al. [103] proposed an unconventional pairing mechanism due to electronic correlations.

CaC_6

The magnetization curve of CaC_6 obtained by Weller et al. [87] does not reach the saturation of the sample due to his low quality. In this study, the authors have used the vapour transport technique described in [26]. This synthesis method is well known to lead in the case of calcium to a weak intercalation in surface. Consequently, Ellerby et al. [104] have estimated the volume fraction of CaC_6 in the measured sample at 3% by X ray diffraction. However, these authors have shown unambiguously that a superconducting state appears in CaC_6 below 11.5 K. Magnetization measurements done on bulk sample [88] conduced to a sharp transition at 11.5 K with a transition width of less than 0.5 K. Field dependence of the magnetic susceptibility was studied with the applied field parallel ($H_{//ab}$) and perpendicular ($H_{//c}$) to the graphene sheets. In both directions, CaC_6 clearly appears as a type II superconductor [88, 105, 106]. The anisotropic ratio

$H_{c2//ab}/H_{c2//c}$ estimated from magnetic measurement is around 2 (Figure 14) with zero field extrapolated values of around 10 kOe and 5 kOe for $H_{c2//ab}$ and $H_{c2//c}$ respectively. Xie et al. have reported higher anisotropy of 3,5-4 [105].

From these results, Csànyi et al. [103] suggest that the relatively high T_c compared to those encountered in this family of compounds can't be explained by a conventional electron-phonon mechanism but by electronic correlations. On the opposite, Mazin [101] proposed a conventional coupling and explained the difference between the T_c of CaC_6 and YbC_6 by the weight difference between Ca and Yb atoms. More recently, Calandra and Mauri [99, 100] attribute the superconductivity of CaC_6 to the coupling of the electrons in the Ca Fermi surface with Ca in-plane and C out-of-plane phonons.

The first experimental indication of the pairing mechanism was obtained from the measurement of the in-plane magnetic penetration depth λ_{ab} [107, 108], using a high-resolution mutual inductance technique [109]. Indeed, the variation $\Delta\lambda_{ab}(T)$ present a thermally activated behaviour, which is compatible with the standard BCS s-wave model.

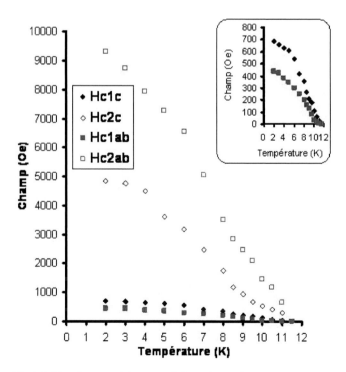

Figure 14. critical fields of CaC_6 with H parallel (H ab) and perpendicular (H c) to the graphene plane.

From this model, the zero-temperature penetration depth was evaluated at $\lambda_{ab}(0) = 72 \pm 8$ nm and the zero-temperature superconducting gap at $\Delta(0) = 1.79 \pm 0.08$ meV. In this case, the ratio $2\Delta(0)/k_BT_c$ is estimated at 3.6 ± 0.2, which is closed to the BCS value of 3.52. Furthermore, this study supports the BCS model of the theoretical work of Calandra and Mauri [99]. In fact, the expected ratio $2\Delta(0)/k_BT_c$ is of 3.69 [110] with the predicted values of electron-phonon coupling $\lambda = 0.83$ and the logarithmic average phonon frequency $\omega_{ln} = 24.7$ meV [99]. Moreover, the thermal behaviour of CaC_6 indicates that this compound is in the dirty limit [107, 108]. More recently, these results were confirmed by the study of the CaC_6 surface resistance using a perturbation method [111, 112]. For $T<T_c/2$, a exponential behaviour of R_s is observed which is the clear signature of a superconducting gap. This comportment can be described in a BCS framework. In this case, the thermal dependence of the surface resistance R_s is proportional to $\exp(-(\Delta(0)/k_BT))$. The gap value $\Delta(0)$ evaluated from this experiment is of 1.7 ± 0.3 meV [111, 112].

The specific heat anomaly measured by Kim et al. [113] has been well fitted by the "α model", according to an isotropic s-wave BCS gap. Its leads to the following values: $T_c = 11.30$ K and $\alpha = \Delta(0)/k_BT_c = 1.776$. This value of α is very close to the weak coupling limit one ($\alpha = 1.76$) [113]. Moreover, the electron-phonon coupling strength was estimated to be $\lambda = 0.70 \pm 0.04$. For the authors, this value is in reasonable agreement with the calculation of Calandra and Mauri [99]. However, Mazin et al [114] pointed out that the above-mentioned electron-phonon theories do not fit well into these data. This problem was solved by recent calculations [115]. The method developed to describe superconductivity in elemental metals produced moderately anisotropic gap function perfectly fitting the specific heat measurements.

Scanning tunnelling microscopy and spectroscopy (STM/STS) [116] indicate a gap value $\Delta(0) = 1.6 \pm 0.2$ meV, in good agreement with the BCS theory. The coherence length in the ab plane, extracted from normalized zero-bias conductance vortex profile, reaches 33 nm, in good agreement with the value obtained from magnetic measurements [88]. The variations compared to the bulk measurements [88, 107, 108, 113] were assigned to the high sensitivity of the sample and of the extremely superficial character of the STM/STS experiments. The authors report also that there is no evidence for a double-gap. A larger gap value $\Delta(0) = 2.3$ meV was reported by Kurter et al [117] suggesting possible strong coupling and anisotropic gap.

A clear proof of this anisotropy was given from directional point contact spectroscopy [118], which confirmed previous STM results and clearly showed the anisotropy of the gap function predicted by theory [115]. However, the gap value is 1.36-1.7 meV, and the anisotropy is probably greater than proposed from first-principles calculations. In addition to the STM experiments, far-infrared reflectance spectra [119] supported the s-wave scenario, confirming the anisotropy of the gap function.

Phonon modes were studied by Raman scattering [120]. Spectra on cleaved surfaces systematically show two bands at 1500 cm^{-1} and 450 cm^{-1} assigned to Raman-active E_g modes originating from the in-plane bond-stretching mode and the out-of-plane puckering one. The in-plane Ca modes might be accessible by far-IR absorption techniques.

An unusual Ca isotope effect coeffcient of 0.52 was reported by Hinks et al [121]. This high value suggests low contribution of the C phonons to the superconducting mechanism. However, small amount of CaC_6 in the samples, due to the intercalation process, could also be a reason, considering the smaller T_c observed in the ^{44}Ca-enriched sample. Nevertheless, recent calculations suggest that the Ca isotopic effect can be explained by the involvement of the Ca $3d$ states [122]. To conclude, as suggest in [100, 123], measurement of C isotope effect and refining the Ca isotope coefficient $\alpha(Ca)$ will be instructive in the comprehension of the superconductivity in CaC_6.

SrC_6

Recently, two other graphite-alkaline earth metal binary compounds were investigated: SrC_6 and BaC_6. Kim et al. have showed that SrC_6 becomes superconductor for temperatures lower than 1.65 K [124] and Nakamae et al. [125] that BaC_6 is not a superconductor at temperatures as low as 80 mK. These results point towards an overestimation of Calandra and Mauri's calculations [126]. Indeed, the autors predicted critical temperatures of 3.0 and 0.2 K respectively.

YbC_6

Weller et al. [87] have demonstrated that YbC_6 is a superconductor with a transition temperature of 6.5 K. This result has been obtained by magnetic measurement on a small sample synthesised by vapour transport process. The X-ray analysis of this sample shows that only 13% of the final volume is made of YbC_6. The magnetic phase diagram presents a relatively small anisotropic ratio $H_{c2//ab}/H_{c2//c}$ of about 2. Concerning the lower critical field H_{c1}, values are quite the same in the both directions.

Like for CaC_6, information concerning the symmetry of the YbC_6 gap is needed. Sutherland et al. [127] report from resistivity and thermal conductivity measurements a s-wave symmetry of the gap function. Moreover, the in-plane coherence length is in the same order of magnitude of the electronic mean free path, that places YbC_6 in the dirty limit.

YbC_6 AND CaC_6 UNDER PRESSURE

Smith et al. [128] have shown that the superconducting transition temperature of YbC_6 increased linearly with an applied hydrostatic pressure in the range 0 – 1.2 GPa, using both resistivity and magnetization measurements. This increase is followed by a maximum of T_c between 7 and 7.1 K at around 1.8 GPa, and then by a subsequent decrease. In the range 0 - 1.2 GPa, the gradient dTc/dP is 0.37 ± 0.01 K/GPa [129]. Moreover, this shift in T_c is reversible. The authors also observed using magnetization measurements that both upper and lower critical fields appear to increase linearly with pressure up to 1.1 GPa. In the case of CaC_6 a similar behaviour is observed in the same range of pressure (0 to 1.2 GPa). The data from magnetization measurements led to a gradient of 0.5 ± 0.05 K/GPa for the variation of T_c with pressure [129]. The positive dependence of T_c on pressure observed for these binaries is the opposite to that observed for ternary mercuro- or thallo-graphitides [130].

This pressure dependence for both YbC_6 and CaC_6 compounds may be explained by a phonon-mediation mechanism but it may also be consistent with a plasmon-mediated superconductivity.

An other investigation on the effect of pressure on CaC_6 up to 1.6 GPa has also shown an almost linearly and reversibly increase of T_c with pressure with a ratio of $(1/T_c)dT_c/dP = 4$ %/GPa [131]. In order to test the electron-phonon mechanism, the authors performed ab-initio calculations on the electronic and vibrational properties of CaC_6. They demonstrated that the behaviour of CaC_6 under pressure may be explained within an electron-phonon mechanism due to a softening of a phonon mode associated to in-plane Ca vibrations. Moreover they predicted that this phonon softening will drive the system unstable for a pressure higher than 12 GPa, and then limit the maximum transition temperature.

This is confirmed by the study of Gauzzi et al. [132]. Indeed, these authors measured the electrical resistivity of CaC_6 at ambient and high pressure up to 16 GPa using conventional four-probe configuration with a dc method. They clearly observed two distinct regimes below and above 8GPa. In the former, the behaviour of CaC_6 is that of a conventional metal with a linear increase of the resistivity with the temperature. The transition temperature also increases linearly with a large rate of 0.5 K/GPa and reaches 15.1 K at 7.5 GPa. The correlation between T_c and $d\rho/dT$ suggests that the T_c increase is caused by a pressure-induced enhancement of the electron-phonon coupling. This phenomenon is studied using first –principles calculations by Zhang et al. [133]. At 8-10 GPa a transition occurred between a good metal with a relatively high T_c and a bad disordered metal with a lower T_c of 5 K. This phenomenon was attributed to the softening of the in-plane Ca phonon mode, leading to structural instability.

X-ray diffraction measurements under pressures up to 13 GPa were carried out to understand this phenomenon [134]. No change in the space group and no extra peaks were observed. However, a clear change in the isothermal compressibility coefficient was detected. Below 9 GPa, da/dP = -0.38 pm/GPa and dc/dP = 8.1 pm/GPa. At 9 GPa, the compressibility of CaC_6 suddenly increases by ~3 times. In addition, a large increase of the XRD linewidth is observed for pressure above 9 GPa revealing pressure-induced disorder. Small displacement of Ca atoms might be responsible for this structural transition, as it was proposed from density functional theory calculation [135]. The associated symmetry reduction might be undetected due to insufficient signal-to-noise ratio.

COMPARISON BETWEEN BOTH LAMELLAR MgB₂ AND CaC₆ SUPERCONDUCTORS

It is interesting to compare boron and carbon, that are two neighbouring elements in the periodic table. Indeed, boron belongs to the 13th column and carbon to the 14th one. And both appear as head of the respective 13th and 14th columns. Consequently, boron possesses three valence electrons and carbon four ones, that are able to create some covalent bonds with other atoms.

When boron undergoes a sp^2 hybridisation, it establishes three simple covalent bonds with three other atoms, and the three bonding angles are exactly equal to 120°. If the three last atoms are identical to the first one (boron sp^2

hybridised), a basic unit is obtained, that, if it is developed to infinity, leads to the formation of an atomic plane made up of boron exactly joined hexagons. Geometrically, this boron plane is quite comparable to a graphene plane (that is itself obtained from carbon atoms, all identical and also sp^2 hybridised). However, both boron and graphene planes exhibit a considerable difference. Carbon belongs indeed to the 14th column, so that its four electrons for three bonding directions only lead to the simultaneous formation of σ- and π-bonds in graphene. On the contrary, the boron plane exhibits exclusively σ-bonds (three electrons and three bonding directions for each atom). The π-orbitals of graphene are of course favourable to the formation of Van der Waals's bonds between successive stacked graphene planes. Indeed, it is here a matter of π-conjugated systems, able to develop a named « π-stacking » structural arrangement. Consequently, the appearance of these weak bonds allows the formation of a solid. Thus graphite originates from these Van der Waals's bonds. On the other hand, when π-orbitals are lacking, these last bonds can't appear. This explain that the hexagonal boron planes are unable to lead to a stacking and consequently to a solid similar to graphite. It doesn't exist in fact an equivalent of graphite obtained from boron atoms.

However, it is possible to think that boron hexagonal planes can exist in more complex materials than an elemental solid. We have seen that graphite exhibits not only a strongly anisotropic structure but also a very anisotropic chemical reactivity. Indeed, using an alkali metal vapour for instance, it is very easy to spread apart the graphene planes, that remain practically unaffected, in order to intercalate the foreign atoms in the dilated spaces. The reaction product is of course a binary graphite intercalation compound, that exhibits mono-layered intercalated sheets ; the metallic atoms occupy some hexagonal prismatic sites, whose apices are defined by carbon atoms. We know that such an intercalation is possible only if an electron exchange takes place between intercalant and adjacent graphene planes. With lithium for instance, we obtain an intercalation compound, whose formula is LiC_6, because the lithium atoms occupy in each intercalated sheet only one-third of the available hexagonal prismatic sites.

Although the « hexagonal boron » doesn't exist, because π-orbitals able to provide the cohesion of the successive stacked planes are lacking, one can think however that well chosen intercalated atoms can promote the stabilization of this structure. In other words, the existence of a hexagonal boron intercalation compound can be considered, even if this boron cannot exist alone. This is observed in fact with magnesium used as intercalant. The MgB_2 compound contains thus hexagonal boron planes, that are systematically separated by mono-layered magnesium planes. The metal atoms occupy of course hexagonal prismatic sites, whose apices are defined by boron ones. In this compound, all prismatic sites are occupied, so that its stoichiometry is especially rich in metal. On the other hand, it has been well established that a noteworthy electron transfer occurs between magnesium atoms and boron planes.

LiC_6 and MgB_2 binary compounds are consequently very similar, if one compares their crystal structures (P6/mmm). Boron hexagons are however very larger than the carbon ones, since the boron-boron distance reaches indeed 216 pm and the carbon-carbon one 142 pm only. Such a difference is explained by the B-B bond index, whose value is 1, while the C-C bond one reaches 1.5. A second reason can explain also this strong difference : indeed the electron transfer appears higher in MgB_2 than in LiC_6. Now it is well known that the injection of additional electrons in a plane of atoms, that are bound one another by covalence, increases systematically the length of this covalent bond.

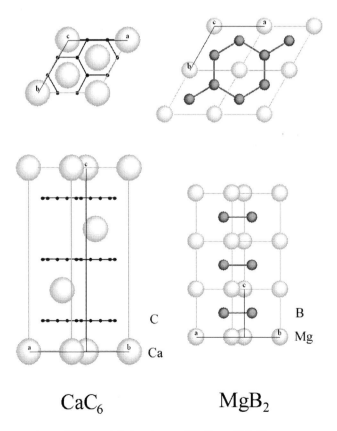

$$CaC_6 \qquad\qquad MgB_2$$

Figure 15. comparison of the crystal structures of CaC_6 and MgB_2.

In MgB_2, the distance between two successive boron planes is equal to 352 pm, since, between two successive carbon ones, it reaches 335, 370 and 452 pm respectively for pristine graphite, LiC_6 and CaC_6 (Figure 15). Both latter values, that are higher than 352 pm, are obviously explained by the presence of π-orbitals in graphene planes. They stand up perpendicularily to the planes and thus generate a space, that is already very large in pristine graphite.

At last, concerning the difference of stoichiometry, one understands that the large size of boron hexagons is favourable to the complete occupation of the prismatic sites, since the small size of the carbon ones favours on the contrary a weaker filling up rate.

MgB_2 appears as a superconducting material with a high transition temperature of 39 K [136]. Specific heat measurements as well as isotopic effect have shown that this material is a type II superconductor consistent with an s-wave BCS model [137]. As for CaC_6, its superconductivity is driven by an electron-phonon interaction. However contrary to CaC_6, STM tunnelling microscopy, point-contact spectroscopy, specific heat measurements and Raman spectroscopy point out the existence of two gaps for MgB_2 [138].

In this compound, the anisotropic ratio of the upper critical field $H_{c2}^{basal\ plane}$ /$H_{c2}^{c\text{-}axis}$ was estimated between 6 and 7, much higher than for CaC_6.

COMPARISON BETWEEN BOTH CaC_6 AND $CaSi_6$ COMPOUNDS

It is interesting as well to compare silicon and carbon, that are also neighbouring in the periodic table. Indeed, calcium is able to react with silicon leading to a binary $CaSi_6$ phase. But its crystal structure is completely different from the CaC_6 one. In fact in a recent paper [139], the authors relate the existence of a metastable $CaSi_6$ compound, which is synthesised using a pressure of 10 GPa and a temperature of 1520 K. When it is put under the ambient conditions, it turns progressively into elemental silicon and $CaSi_2$ usual silicide. Although silicon and carbon belong to the same 14th column of the periodic table, the first one makes use of single bonds only, while with carbon it is usual to observe single, double and even triple bonds. In CaC_6, the carbon atoms, which are sp^2 hybridised, lead to 2D graphene planes containing simultaneously σ- and π-bonds. These latter are delocalized in every successive planes. Calcium atoms intercalate between the planes after spreading apart these ones, so that their coordination by carbon is 6. On the contrary, in $CaSi_6$, the sp^3 hybridised silicon atoms are chained up themselves by single bonds, leading to a 3D clathrate-like framework which contains hexagonal cages, where are embedded calcium atoms with a coordination by silicon of 18. In both cases, calcium is ionised and the excess electrons are found in the basins of C-C or Si-Si covalent bonds. Concerning the electronic properties, they are found very different, since CaC_6 appears as a metal and $CaSi_6$ as a very bad conducting material. On the other hand, CaC_6 becomes superconducting at low temperatures, while $CaSi_6$ is diamagnetic.

TERNARY GRAPHITE INTERCALATION COMPOUNDS

The ternary electron donors graphite intercalation compounds contain necessarily an alkali metal. This latter supplies the indispensable reducing feature, which allows the intercalation reaction. But, on the other hand, the second intercalated element can be as well electropositive as electronegative. It was well established that this second element is in a very wide electronegativity range. Indeed, it can be a second alkali metal or, on the contrary, an halogen or oxygen.

The simplest case is observed, when both intercalated elements are alkali metals. They exhibit very close electronegativities. Consequently, the intercalated sheets of these ternaries appear as mono-atomic layers, as for the binary phases. Instead of a metallic plane into the graphitic galleries, it is a disordered 2D substitution alloy, which is intercalated between the graphene layers. Sodium, potassium, rubidium and cesium [15-17] lead in pairs to $M_xM'_{1-x}C_8$ first stage ternary compounds, for which the 2D hexagonal unit cell is exactly identical to that of KC_8 or CsC_8. The solid solution is continuous for all values of x, included between 0 and 1. The interplanar distance of such a ternary phase is intermediate between that of both MC_8 and $M'C_8$ compounds, but it does not obey the Vegard's law. Among the alkali metals, lithium is unable to give such ternary compounds [27], because of its too small size. It is indeed incompatible with that of the other alkali metals.

When the difference between the electronegativities of both intercalated elements becomes sufficiently large, it is no more possible to obtain mono-atomic intercalated layers. A novel *c*-axis stacking becomes then more stable, with poly-layered intercalated sheets, mostly three-layered. Belonging to the 12th, 13th and 15th columns of the periodic table, mercury, thallium, bismuth, antimony and arsenic are metals or semi-metals, which were thus intercalated into the graphitic galleries in the company of heavy alkali metals. On the whole, with a middling electronegativity, these elements make up the central part of the intercalated sheets. It is surrounded by two alkali metal layers, which are on contact with the graphene planes.

With mercury, the intercalated sheets is three-layered [140, 141], according to the c-axis K-Hg-K or Rb-Hg-Rb sequence (Cs does not intercalate in association with mercury). In the case of potassium, two first stage phases were reported. The pure α compound (pink) is very easily to obtain, while the β phase (pale yellow) is observed only in mixture with the previous one. For rubidium on the other hand, only the pink α phase is known. The second stage compounds (blue) are exclusively reported for the α variety.

In all α phases, the graphene plane and the adjacent potassium (or rubidium) layer are commensurate, so that the C/alkali metal ratio is equal to 4, as in MC_8 binary compounds. In the graphitic gallery, two intercalated K (or Rb) layers are exactly superimposed, leading to triangular prismatic sites, which are all occupied by a mercury atom. Consequently, the 2D unit cell of these intercalated sheets are of course identical to that of MC_8 binary phases. Thus, the chemical formulas of these ternaries have to be written $MHgC_4$ for a first stage compound and $MHgC_8$ for a second stage one. It is interesting to consider the c-axis stacking corresponding to these ternaries, since it appears a perfectly alternated sequence of electronegative and electropositive atomic planes, according to the following succession : C^- K^+ Hg^- K^+ C^-. Both first stage $KHgC_4$ and $RbHgC_4$ compounds were well characterized by the determination of their 3D unit cells, which are orthorhombic (Fddd space group, Figure 16). Their c parameter is four times as large as their repeat distance (1016 pm for $KHgC_4$ and 1076 for $RbHgC_4$), as it is clearly indicated on the figure. The 3D structure of the second stage alkali metal amalgams graphite intercalation compounds has never been reported.

The β potassium amalgam first stage compound shows a larger repeat distance, since it reaches 1076 pm instead of 1016 pm, even though the mercury amount of its intercalated sheets is considerably weaker, leading to the following chemical formula : $KHg_{0.5}C_4$.

In the case of Tl, we note that potassium- and rubidium-thallium graphite intercalation compounds [142, 143] possess an unexpected and more complicated structure (cesium-thallium alloys do not intercalate). Indeed, the usual phases contain five-layered intercalated sheets, leading to especially thick interplanar distances and to very large metal concentrations. We know thus the following compounds : $MTl_{1.5}C_4$ (first stage) and $MTl_{1.5}C_8$ (second stage), with M = K and Rb. For the first stage $KTl_{1.5}C_4$ compound, the repeat distance reaches 1210 pm.

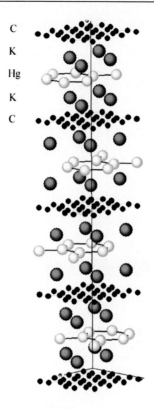

Figure 16. crystal structure of $KHgC_4$ (Fddd).

In the case of rubidium, two different first stage compounds are known with the following repeat distances : 1265 pm (α phase) and 1340 pm (β phase). In these ternary phases, the central part of the intercalated sheet is made up of three thallium layers, and it is surrounded by two alkali metal layers, according to the **c**-axis M-Tl-Tl-Tl-M sequence (Figure 17).

Nevertheless, a K-Tl-K three-layered sheet was reported in the case of the $KTl_{0.67}C_4$ compound (γ phase), whose repeat distance reaches only 1076 pm. The crystal structures of the alkali metal-thallium ternary compounds remain currently imperfectly known. A square 2D unit cell was suggested however (no-commensurate of course) for $KTl_{1.5}C_4$, with a parameter of 1099.5 pm.

It is well established that intercalating lithium with a second element shows very great difficulties [27]. Nevertheless, two metals (only) were reported as able to intercalate into graphite associated with lithium : calcium [14, 144] and europium [145]. In spite of very similar electronegativities, Li and Ca (or Li and

Eu) simultaneously penetrate into graphite leading to poly-layered intercalated sheets.

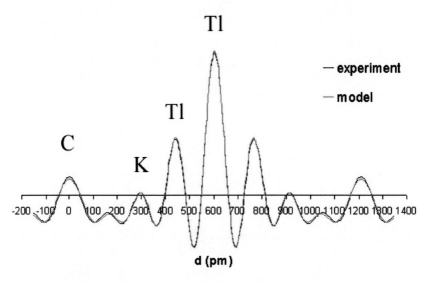

Figure 17. *c*-axis electronic density profiles of $KTl_{1.5}C_4$.

The α first stage graphite-lithium-calcium compound ($Li_{0.5}Ca_{2.8}C_6$) exhibits a *c*-axis interplanar distance of 776 pm and a five-layered intercalated sheet which has something in common with a Li-Ca-Li-Ca-Li slice cut in the $CaLi_2$ structure (ThMn$_2$ Laves phase type). The β phase (first stage) is richer in metallic elements ($Li_3Ca_2C_6$). It possesses a greater repeat distance (970 pm) and a seven-layered intercalated sheet due to the splitting of the middle lithium plane in three according to the Li-Ca-Li-Li-Li-Ca-Li sequence : the central lithium layer is very rich in metal (1.5 lithium atoms for every six carbon atoms), whereas the outer lithium planes (130 pm above and below) contain only 0.6 lithium atoms for every six carbon atoms (Figure 18).

A first stage ternary graphite-lithium-europium compound has been also reported. Its *c*-axis interplanar distance reaches 804 pm. As previously, the intercalated Li-Eu alloy is made up of several superimposed atomic layers.

Associated with heavy alkali metals, bismuth [146, 147], antimony [148, 149] and arsenic [150], which belong all to the 15th column of the periodic table, give rise to ternary graphite intercalation compounds. All these ternaries possess three-layered intercalated sheets, so that their interplanar distances remain rather weak and included between 950 pm (potassium-arsenic-graphite) and 1150 pm (cesium-bismuth-graphite).

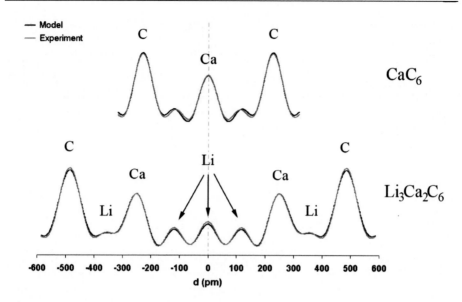

Figure 18. c-axis electronic density profiles of CaC_6 and $Li_3Ca_2C_6$.

Their 3D crystal structures are not known, but, in most cases, their in-plane unit cells were reported : they can be commensurate or not with 2D graphene lattice, hexagonal or not (rectangular, oblique), according to the cases. On the other hand, it exists a very important polymorphism for the Sb and As ternary compounds, so that these systems lead to especially complex lists of phases. The situation is however a little simpler in the cases of Bi-K, Bi-Rb and Bi-Cs ternary graphite intercalation compounds. Indeed, stages 2 to 5 ternary compounds were observed in the case of Bi-K intercalated alloys, with several interplanar distances which vary between 987 and 1086 pm. On the other hand, stages 1 and 2 only are obtained with Bi-Rb and Bi-Cs intercalated alloys, exhibiting several interplanar distances included between 1009 and 1052 pm and between 1060 and 1150 pm for rubidium and cesium respectively. The in-plane structures of these Bi-heavy alkali metal graphite intercalation compounds exhibit either hexagonal or oblique symmetry, according to the cases.

Phosphorus belongs also to the 15th column of the periodic table, as As, Sb and Bi, but it is however clearly different because of its stronger electronegativity (2.1 in the Pauling's scale), which forbid it to behave as a semi-metal. A first stage blue $KP_{0.3}C_{3.2}$ ternary compound was nevertheless reported, with three-layered K-P-K intercalated sheets and a repeat distance of 886 pm [151]. It is clear that the high electronegativity of phosphorus promotes the formation of this structure. On the other hand, this ternary compound appears as greatly stable against air and water.

The electronegativities of hydrogen and phosphorus are very close, so that it is not surprising that hydrogen leads also, associated with alkali metals, to ternary graphite intercalation compounds. Two synthesis methods are recommended in order to prepare these ternaries. The first one [152-155] consists in introducing hydrogen atmosphere in the presence of KC_8 or RbC_8 (with CsC_8, no reaction occurs at all), using possibly high pressures. The hydrogen absorption by these first stage binary phases leads to second stage $KH_{0.67}C_8$ or $RbH_{0.67}C_8$ ternary compounds. Simultaneously, the alkali metal intercalated mono-layers turn into M-H-M three-layered sheets. Of course, the interplanar distances considerably increase during the hydrogenation : from 535 to 853 pm in the case of potassium and from 565 to 903 pm in the case of rubidium. The second synthesis method is a direct reaction between alkali metal hydride and graphite [156-158]. Its advantage lies in the opportunity to prepare also first stage ternary compounds, and moreover it allows to obtain ternaries with almost all alkali metals (from sodium to cesium). The formulas of the first stage compounds can be written as following : $NaHC_{2.5-4.5}$, $KH_{0.8}C_4$, $RbH_{0.8}C_{3.6}$ and $CsH_{0.8}C_5$. With NaH, the value of the interplanar distance is 730-760 pm and it reaches 1005 pm in the case of CsH. All these ternaries include obviously three-layered intercalated sheets, whose central layer is made up of hydride anions. In these ternaries, the 2D structure of the intercalated sheets is mostly orthorhombic [159].

Lastly, it is possible to build three-layered intercalated sheets, using an alkali metal associated with a strongly electronegative element as chalcogen or halogen. The most well known among the corresponding ternaries are the oxygen-alkali metal graphite intercalation compounds. They are prepared by heat treatment-induced direct reaction of pyrolytic graphite with partly oxidized liquid alkali metal, which contains very small amounts of peroxide or hyperoxide anions. Most of these ternaries are stable in air.

In the case of sodium, an unexpected stage 2 blue compound is observed [160]. Its chemical formula is $NaO_{0.44}C_{5.9}$ and its intercalated sheets are five-layered according to the Na-O-Na-O-Na c-axis stacking, so that the interplanar distance reaches 745 pm. This unusual sequence can be explained by the presence of O_2^{2-} peroxide anions ; indeed, it has been well established that the intercalated sheets are simply 2D slices of Na_2O_2 sodium peroxide with small distortion, due to the intercalation. The 2D unit cell is hexagonal with a parameter of 636 pm, and their is tilted of $\pm 0.7°$ with respect to the graphenic one.

First stage oxygen-potassium graphite intercalation compounds of formula close to $KO_{0.07}C_4$ were reported as quasi-binary compounds, with repeat distances included between 840 and 890 pm, depending on different polymorphs [161, 162]. Their intercalated sheets are two-layered (K-K), but the electrostatic repulsion between both cationic layers is deleted by the intermediate presence of a very small amount of O^{2-} anions. Isostructural ternaries with Rb and Cs were also obtained [163, 164], although the oxygen concentration is higher ($RbO_{0.1-0.3}$ C_4 and $CsO_{0.3-0.6}$ C_4).

Using reagents containing very large amount of oxygen, it is possible to synthesize especially rich in oxygen ternaries. Thus, the reaction between KC_8 and KO_2 hyperoxide provides KO_2C_4 (first stage) and KO_2C_8 (second stage) ternaries [165], whose interplanar distance is 844 pm. And the reaction of Cs_3O with graphite leads to a first stage ternary [166], whose formula is $CsO_{1.2}C_8$ and repeat distance 984 pm. Because of the large amount of oxygen included in their intercalated sheets, these latter are four-layered according to the M-O-O-M c-axis sequence. As usually, the cations/anions alternation in this sandwiched structure is favourable to its stabilization.

As oxygen, sulfur, selenium and tellurium behave to the 16th column of the periodic table. These chalcogens are also able to intercalate into graphite in association with an alkali metal, although their electronegativities are considerably lower than that of oxygen. The corresponding potassium ternaries, which were especially studied, are prepared by immersing a pyrolytic graphite platelet in liquid potassium containing a small amount of dissolved chalcogen [167, 168]. Purple first stage compounds are obtained, which are very stable against air and water and exhibit the following compositions : $K_2S_{0.5}C_6$ and $K_2Se_{0.2}C_6$. Their c-axis repeat distances are 875 and 871 pm respectively. With three-layered K-S-K or K-Se-K intercalated sheets, these ternary phases possess in-plane structure, which are commensurate with the graphenic one. Indeed, the parameter of the 2D hexagonal unit cell is equal to 426 pm for $K_2S_{0.5}C_6$. The sulfur atoms, located in the central layer of the sandwich, occupy the centers of an octahedral environment of potassium atoms, so that the intercalated sandwich appears clearly as a 2D slice of the fcc antifluorite K_2S potassium sulfide. On the other hand, the sulfur atoms statistically occupy one-half of the octahedral sites, leading to the S/K atomic ratio of ¼ (Figure 19).

Alkali metal halides were reported to give also donor-type ternary graphite intercalation compounds. The molten alkali metal in presence of its halide reacts indeed with graphite in a stainless steel tube under an argon atmosphere [169]. The case of sodium halides is the best known. It was reported for NaCl, NaBr and NaI [170]. The corresponding ternaries, with intercalated sheets, whose thicknesses are included in the range 700-800 pm, possess three-layered Na-X-Na (X = halogen) intercalated sandwiches in the graphitic galleries. For Cl and Br respectively, the interplanar distances reach 756 and 771 pm. For iodine, the system is more complex, since four different second stage ternaries were observed.

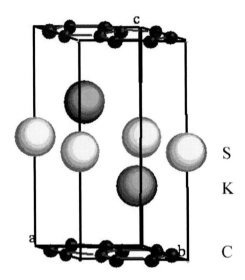

Figure 19. crystal structure of $KS_{0.25}C_3$ (P-3m1).

We have seen that almost all the donor-type ternaries possess poly-layered intercalated sheets (and even three-layered ones generally). The layer of more electronegative species is sandwiched by two layers of electropositive species (alkali metal). And it is possible to distinguish two types of ternaries according to the electronegativity of the sandwiched element: alloy intercalates (case of mercury, for instance) and ionic intercalates (case of oxygen).

The intercalated alloy makes a large contribution to the electronic structure around the Fermi level. KHg potassium amalgam, which has been especially studied, appears as a typical intercalated alloy [171, 172]. For the corresponding ternaries, the electronic structure of the intercalate consists of K 4s and Hg 6s-6p levels in the region close to the Fermi energy, which is superimposed upon the graphitic π-bands (Figure 20). The presence of the Hg-inherited state around the Fermi energy seems to play an important role in the appearance of the superconductivity in $KHgC_4$ and $KHgC_8$.

On the other hand, potassium-hydrogen and potassium-oxygen appear as characteristic of ionic intercalates. Owing to the charge transfer from potassium to graphite and hydrogen, the H atoms in $KH_{0.8}C_4$ and $KH_{0.8}C_8$ are negatively charged by reason of the electronegativity of this element.

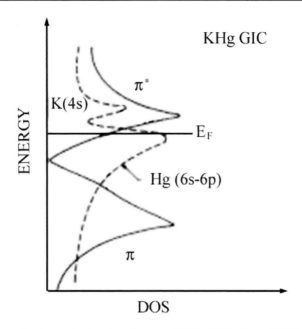

Figure 20. schematic electronic density of states of KHg GICs (from [172]).

These H⁻ ions are stabilized by the presence of both K⁺ cations layers, that surround the central hydrogen layer [154, 173]. By proton-NMR measurements [162], the metallic nature of the hydrogen species has been established in these ternaries, so that a 2D metallic hydrogen lattice is obtained inside the graphitic galleries. This novel result was also predicted by band calculations [175]. In the intercalated sheet, the nearest hydrogen-hydrogen distance ranges around 296 pm, which is slightly smaller than the diameter of the H⁻ anion (308 pm). Thus, the H-H distance reaches a value as short as the distance observed in metallic hydrogen, which can be obtained by application of very high pressures [176].

For the potassium-oxygen KO_2C_4 and KO_2C_8 ternaries, a band calculation was carried out on the basis of the local density functional formalisms [177]. The intercalate band appears as superimposed upon the graphitic conduction π*-band, so that the oxygen double layers contained in the intercalated sheet possess also a weakly metallic character.

Finally, the electronic structure of the $K_2S_{0.5}C_6$ ternary compound was studied by means of a first-principles density functional theory approach [178]. The analysis of the band structure and density of states (DOS) shows that both graphene and K layers provide partially filled bands and thus contribute to the metallic conductivity of the material. The calculated Fermi surface is reported in Figure 21, which, for clarity, shows a view along a direction slightly tilted with respect to the interlayer c-axis direction. The K sublayers of the intercalate were shown to have an important contribution to the Fermi surface.

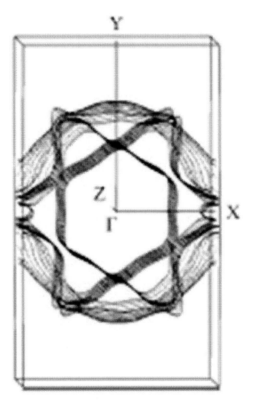

Figure 21. Fermi surface of $KS_{0.25}C_3$ (from [178]).

We have seen that, concerning their conductivity, the binary electron donors graphite intercalation compounds exhibit a partial three-dimensionality, as it was well observed examining the room temperature σ_a/σ_c ratio. This characteristic three-dimensionality of conductivity completely disappears in the case of the ternary electron donors graphite intercalation compounds. Indeed, numerous experiments concerning the measurement of the room temperature σ_a/σ_c ratio of such ternary compounds show their strong two-dimensional character [179]. It reaches, in most cases, 10^4-10^5, so that, concerning this specific property, these ternary compounds appear as very similar to the electron acceptors ones.

Chapter 8

SUPERCONDUCTIVITY OF TERNARY GRAPHITE INTERCALATION COMPOUNDS

Among donor type graphite intercalation compounds, some ternaries containing an alkali metal associated with a less electropositive element as mercury, thallium and bismuth exhibit superconducting properties (Table II). Contrary to the binary compounds, we have seen that their intercalated sheets are poly-layered.

Table II. Superconducting ternary GICs

GIC	I_c (pm)	Stage	$H_{c2}^{\perp}/H_{c2}^{//}$	T_c (K)	References
KHgC$_4$	1015	1	11	0.73	184, 191
	1024 (α-phase)			1.5	185
	1083 (β-phase)			0.7	185
	1020			1.4	186
RbHgC$_4$	1076	1	10	0.99	184, 191
KHgC$_8$	1350	2		1.93	180
			30 - 31	1.70 – 1.90	182
			40	1.90	181
			15.5	1.90	184, 191
			21.7	1.94	190
			25	1.90	192
RbHgC$_8$		2		1.46	183
				1.44	181
	1411		20 - 40	1.40	184, 191
KHgC$_{12}$		3		1.94	187
KTl$_{1.5}$C$_4$	1208	1		2.7	189
			10	2.56	191
			5	2.7	190
KTl$_{1.5}$C$_8$	1548	2		2.45	189

Table II. Continued

				1.25 – 1.5	193
RbBi$_{0.6}$C$_8$ α		2		1.5	18
CsBi$_{0.5}$C$_4$ α	1070	1		4.05	146, 193, 18
CsBiC$_4$ β		1		2.3	193,18
CsBiC$_8$ β		2		2.7	193
				3.45	18
Cs – Bi α,β		1		4.69	194
Cs – Bi α,β		2		4.72	194
Cs – Bi β		2		4.73	194
Cs – Bi β		3		4.72	184
Li$_3$Ca$_2$C$_6$	970	1	1.5	11.15	190

ALKALI METAL AMALGAMS INTERCALATED COMPOUNDS

The superconductivity in this family of compounds was suspected in 1980 by Alexander et al. [180] when they observed an anomaly in the low temperature specific heat at 1.93 K in the second stage KHgC$_8$ compound. It was then clearly observed in 1981 by Pendrys et al. [181] by AC susceptibility measurements using a contactless induction technique. The superconductivity of this compound was confirmed by Koike and Tanuma [182] by AC magnetic susceptibility measurements too. They found critical temperatures of 1.90 K and 1.70 K for two different samples. Superconductivity was also observed in RbHgC$_8$ with a T$_c$ of 1.46 K obtained using specific heat measurements [183] and 1.44 K by AC induction technique [181].

Iye et Tanuma [184] confirmed the critical temperature of 1.90 K and 1.40 K for KHgC$_8$ and RbHgC$_8$ and revealed the superconductivity of KHgC$_4$ and RbHgC$_4$ under 0.73 K and 0.99 K respectively. They compared these transition temperatures with those of the free amalgams. Indeed, contrary to the superconducting binary GICs in which the intercalate is not superconducting, in the case of ternaries containing alkali metals amalgams, the latter are superconducting with critical temperatures of 0.94 K and 1.20 K for KHg and KHg$_2$ respectively, and 1.17 K for RbHg.

Transition temperatures between 0.7 K and 1.5 K are reported more recently by Chaiken et al. [185] for KHgC$_4$. The lowest T$_c$ are obtained for samples containing the β-phase (repeat distance of 1083 pm) and the highest T$_c$ are measured for samples of pure α-phase (repeat distance of 1024 pm).

Superconducting graphite fiber of $KHgC_4$ with a transition temperature of 1.4 K was also successfully synthesised by Tsukamoto et al. [186] from a highly graphitized graphite fiber (150 μm diameter) prepared by CVD.

Moreover, the stage 3 potassium-mercury compound appeared to be also superconducting, with a T_c of 1.94 K, the highest among these compounds [187].

The hydrogenation of first stage mercurographitides leads to superconducting compounds with a critical temperature of 1.5 K [188].

All compounds are type II superconductors. The critical field anisotropy ratio $H_{c2}^{\perp}/H_{c2}^{//}$ is about 10 for the stage 1 and it ranges from 15 to 40 for the stage 2 compounds. It is argued that the electrons in the intercalate band rather than those of the graphitic ones play the main role in the superconductivity. It is also interesting to underline the fact that the stage 2 compounds that have a lower density of states at the Fermi level present a higher transition temperature than the stage 1 compounds.

ALKALI METAL – THALLIUM ALLOYS INTERCALATED COMPOUNDS

Higher critical temperatures of 2.7 K and 2.45 K were observed in ternary graphite intercalation compounds containing alkali metal alloyed with thallium : the first stage $KTl_{1.5}C_4$ and the second stage $KTl_{1.5}C_8$ respectively [189]. Before the recent discovery of the superconducting properties of the $Li_3Ca_2C_6$ ternary compound [190], $KTl_{1.5}C_4$ had the highest value of T_c observed among the ternary graphite intercalation compounds.

ORIGIN OF SUPERCONDUCTIVITY

The origin of superconductivity is discussed for mercurographitides and thallographitides [181]. As the transition temperatures of alkali metal amalgams are very close to those of GICs containing KHg and RbHg alloys, the electrons in the intercalate band rather than those in the graphitic band would play a major role in the superconductivity. Moreover, the second stage KHg compound exhibits a larger anisotropy of the critical fields than the first stage compound, so that the superconductivity is associated with intercalate layers rather than graphene planes since the two-dimensionality increased with the stage number. In the same way, $KTl_{1.5}C_4$ that is richer in metal than the previous compounds and presents a higher

T_c among this family of ternaries suggests also the leading role of the metal layers. In all compounds, the critical field anisotropy is well explained by the effective mass model. Even for compounds belonging to stage 2, the **c**-axis coherence length remains higher than the repeat distance. Whereas these compounds are highly anisotropic, their superconductivity is essentially three-dimensional.

Pendrys et al. [180] found a nearly temperature independent anisotropy of the highest critical field H_{c2} ($H_{c2}^{\perp}/H_{c2}^{//}$) of 5 in $KTl_{1.5}C_4$ whereas the extrapolated value at T_c for $KHgC_8$ compound is around 25 and appears to increase at reduced temperatures. They concluded that their superconducting behaviour is well described by anisotropic three-dimensional models and that the superconductivity is mainly dominated by the electron-phonon interactions.

APPLICATION OF PRESSURE

The application of pressure [191] up to 6 kbars on $KHgC_8$ and $RbHgC_8$ led to a large increase of the c-axis coherence length with pressure and a small increase of the coherence length in the perpendicular plane. Consequently the anisotropy clearly decreases with the application of pressure. In the case of $KTl_{1.5}C_4$, that is less anisotropic at ambient pressure, this phenomenon is less pronounced. For these compounds, the critical temperature decreases linearly with the applied pressure and the pressure coefficients dT_c/dp are of the same order of magnitude as those of superconductors such as tin or mercury.

ALKALI METALS – BISMUTH TERNARY COMPOUNDS

In 1985, Lagrange et al. revealed the superconductivity of several compounds among a new family of ternaries : the graphite-alkali metal-bismuth compounds [18, 134, 193] by means of resistivity measurements. The authors reported the superconductivity of stage 2 rubidium–bismuth compound and cesium-bismuth compounds belonging to stage 1 and 2. The α graphite-cesium-bismuth compound of stage 1 exhibits the highest value of Tc : 4.05 K. This results has been observed in several samples, however it happened that several compounds prepared using the same synthesis method didn't show the same transition temperature. Stang et al. [194] studied cesium-bismuth compounds of stage 1 to 3 using AC inductance method. Superconducting compounds were observed when they were synthesised from starting alloys $Cs_{1-x} - Bi_x$ with x higher than 0.5. In this case, the transition temperature is very close to that of the intermetallic compound $CsBi_2$ ($T_c = 4.75$ K) [195] and the superconducting volume fraction is between 0.1 % and 5 %. Consequently, the presence of small quantities of the binary alloy as inclusions in the GICs are suspected to be at the origin of the superconductivity.

RECENT PROGRESS IN $Li_3Ca_2C_6$

A recent work on the graphite-lithium-calcium system revealed the superconductivity of the β phase of this system : $Li_3Ca_2C_6$ with a transition temperature of 11.15 K [190]. As for CaC_6 among the binaries, it represents an increase of T_c by one order of magnitude over all graphite intercalation compounds.

This compound is very rich in metal since its intercalated sheets are seven-layered according to the following sequence : Li-Ca-Li-Li-Li-Ca-Li with a repeat distance of 970 pm [14].

Using magnetization measurements, a clear diamagnetic transition appeared at 11.15 K, however, the saturation is not reached down to 2 K. This is probably due to the complexity of the building of the compound that creates some defects. The field dependence of the magnetization was investigated at different temperatures in order to determine the magnetic phase diagram with the applied field parallel or perpendicular to the c-axis. The critical fields and the superconducting volume fraction were estimated from magnetization measurements versus applied field at several temperatures. The superconducting volume fraction depends on the sample but was evaluated for the best samples at 55 ± 5 %. $Li_3Ca_2C_6$ presents a type II superconducting behavior. The lower critical fields H_{c1} are roughly the same in both directions (with H applied parallel ($H_{c1//c}$) and perpendicular ($H_{c1//ab}$) to the c-axis) with an extrapolated value at T = 0 K of 30 Oe. The upper critical fields show a sizable anisotropic ratio $H_{c2//ab}/H_{c2//c}$ of 1.5 with extrapolated values at 0 K of 3100 Oe and 2000 Oe for $H_{c2//ab}$ and $H_{c2//c}$ respectively. These higher critical fields are roughly three times lower than those of CaC_6. In spite of their lamellar structure, both superconducting compounds belonging to the graphite-lithium-calcium system, CaC_6 and $Li_3Ca_2C_6$ exhibit an anisotropic three-dimensional superconducting behavior.

Chapter 9

SHORT GENERAL IDEA ABOUT THE SUPERCONDUCTIVITY OF THE OTHER INTERCALATED CARBON MATERIALS

Among carbon materials, compounds derived from the two-dimensional allotropic variety are not the own to exhibit superconductive properties. Indeed, diamond (3D host lattice), carbon nanotubes (1D) and fullerite (0D) lead also to superconducting materials.

DOPED-DIAMOND

Diamond is a electrical insulator, however boron is able to act as a charge acceptor towards diamond, the resulting diamond is hole-doped. This material is synthesised by reaction between graphite and B_4C under high pressure (8-9 GPa) and high temperature (2500-2800 K) for a very short time (5 s) [196]. Electrical resistivity, magnetic susceptibility, specific heat measurements show that this boron-doped diamond is bulk-type II superconductor below a transition temperature of 4 K.

CARBON NANOTUBES

The superconductivity in carbon nanotubes is still the argument of wide discussion and the object of several theoretical and experimental works. Superconductivity has been reported in several specific cases. It can be induced at 4.2 K by a proximity effect in single-walled carbon nanotubes connected to niobium electrodes [197]. Proximity-induced superconductivity in single-walled carbon nanotubes (SWNT) below 1 K, both in a single tube 1 nm in diameter and in ropes containing. about 100 nanotubes, was also observed [198]. In this experiment, the samples were suspended between two superconducting electrodes, permitting structural study in a TEM microscope. When the resistance of the nanotube junction is sufficiently low, it becomes superconducting and can carry high supercurrents.

Kociak et al. [199] reported superconductivity under 1 K in ropes of SWNT in contact with sputtered $Al_2O_3/Pt/Au$. The tubes are not doped and the authors argued that the superconducting transitions are not due to a proximity effect.

Single-walled small-diameter carbon nanotubes (400 pm) embedded in a zeolite matrix exhibit a superconducting behaviour as an anisotropic Meissner effect with a transition temperature of 15 K [200].

Very recently, Takesue et al. [201] reported superconductivity at 12 K in entirely end-bonded multi-walled carbon nanotubes (MWNT) in Au/MWNTs/Al junctions prepared in nanopores of alumina templates.

The common feature among these different cases is that the superconducting state appears only when an increase in the density of charge carrier occurs.

FULLERIDES

Superconductivity was observed for the first time in 1991 by Hebard et al. [202] in alkali metals intercalated C_{60}. Many experiments have been done in the ninety's, drawn from the data of graphite intercalation compounds. However, in spite of the discovery of the superconductivity in CaC_6 and in $Li_3Ca_2C_6$, the transition temperatures of the C_{60} compounds are much higher (up to 40 K for Cs_3C_{60} under 12 kbar) than those of GICs. In the latter, in spite of the structure of the host lattice and some anisotropic properties the superconductivity is not that of a two-dimensional system since the coherence lengths are large compared with the repeat distances. C_{60} compounds are characterized by an isotropic structure and then a three-dimensional behaviour. However, the differences in both electronic structure and phonon system between GICs and fullerides are sufficiently important to contribute to the difference in superconductivity [5]. An early review of superconducting fullerides was made by Lüders [195] in 1992.

The first superconducting compounds discovered were the binary alkali metal fullerides : K_3C_{60} and Rb_3C_{60} with transition temperatures of 18 K [202] and 30 K respectively [203]. Contrary to the GICs, the intercalation of alkali metals alloys into fullerite led to superconducting materials.

And it was early demonstrated that among the metallic "M_3C_{60}" compounds, the higher the a parameter of the cubic lattice, the higher the T_c [204]. K_2RbC_{60} and KRb_2C_{60} exhibit superconducting properties under 21.8 and 26.4 respectively [205], whereas Rb_2CsC_{60} and $RbCs_2C_{60}$ present higher T_c of 31 K and 33 K [206]. Cs_3C_{60} isn't stable in ambient conditions [207], but it becomes superconducting at 40 K under pressure [208].

The superconductivity of ternary compounds containing lithium or sodium associated with a heavy alkali metal was studied in detail by Tanigaki et al [209]. Na_2CsC_{60} and Na_2RbC_{60} are superconducting at 12 K and 3.5 K, while Li_2MC_{60} (M = K, Rb, Cs) present no superconductivity down to 2 K. An overview of the variation of the transition temperature with respect to the lattice parameter for various fullerides containing alkali metals, given by Margadonna and Prassides [210] is depicted in Figure 22.

An important issue in fullerides superconductivity, related to the correlation between T_c and interfullerene separation, was the search for new materials with larger lattice parameters. An excellent method to reach this aim is to solvate alkali ions with neutral molecules such as ammonia. It was succefully realised by Zhou et al. [211] with Na_2CsC_{60} : the reaction with ammonia led to $(NH_3)_4 Na_2CsC_{60}$ that became superconducting at 29.6 K.

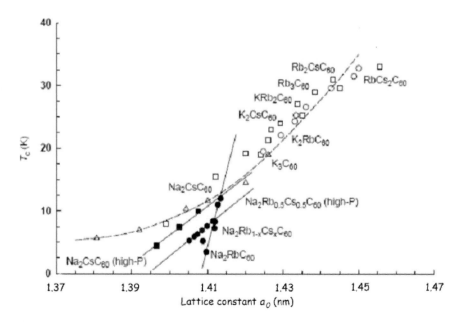

Figure 22. evolution of T_c with lattice constant of alkali metals fullerides (from [210]).

Several attempts were done in order to synthesise ternary compounds using "nominal compositions" like "$(KTl_{1.5})_3 C_{60}$" in strong correlation with superconducting GICs Transition temperatures were determined but when all results were correlated, it appeared that these compounds were not ternary doped fullerene [195] but generally only slightly modified binary compounds.

Alkaline earth fullerides were also studied. Early work gave the structure of the first compound of this family : Ca_5C_{60} that become superconducting under 8.4 K [212]. In the case of barium, the first work showed a superconducting phase the chemical formula of which is Ba_6C_{60} with a transition temperature at 7 K [213]. But this result was discussed and the bulk superconducting phases in the Ba- and in the Sr-C_{60} systems have been established unambiguously for Ba_4C_{60} and Sr_4C_{60} with T_c of 6.7 K and 4.4 K respectively [210].

The $A_3Ba_3C_{60}$ (A = K, Rb, Cs) was also studied. Indeed, bulk superconductivity was observed for $K_3Ba_3C_{60}$ with a T_c of 5.6 K and $Rb_3Ba_3C_{60}$ (T_c = 2 K) while $Cs_3Ba_3C_{60}$ is not superconducting down to 0.5 K [210].

The superconductivity of rare earth fullerides is still a controversial topic. Indeed, it was reported in 1995 by Ozdas et al. [213] and Chen and Roth [214] for $Yb_{2.75}C_{60}$ and $Sm_{2.75}C_{60}$ with transition temperatures of 6 K and 8 K respectively. Recently Akada et al. [215] claimed that both compounds are not superconducting and only $Yb_{3.5}C_{60}$ exhibit a superconducting behaviour under 6 K.

Chapter 10

CONCLUSION

The electron donors graphite intercalation compounds appear, in the field of the solid state chemistry, as a family of lamellar materials, which, for several of them, become superconducting at low temperatures.

Concerning their electronic properties, these compounds possess, in the normal state, metallic features, with a pronounced anisotropy. The latter is quite small in the case of the binaries, especially when the intercalant little spreads apart the graphene planes. But it can, on the contrary, become extremely high for the ternary compounds, which exhibit a very large space between the carbon layers. In all cases, it is possible to consider as quasi-metallic their in plane conductivity, but the c-axis one, inevitably weaker, is more difficult to represent. The charge transfer from the intercalant to the graphene layers is systematic ; it releases electrons into these planes, whose conductivity is similar to the metallic one.

It is not easy to understand why several compounds are able to become superconducting at low temperatures, while the others keep their normal state. Among the superconducting phases, we find as well binaries as ternaries, without correlations as regards their critical temperatures.

However, it is interesting to note that the charge transfer seems to play an important role for the appearance of the superconductivity in these materials. One can say that higher is the charge transfer, higher is the corresponding critical temperature.

Indeed, it was well established that both ytterbium and calcium generate very large charge transfers during their intercalation into graphite, leading to noteworthy dilations of the graphene planes. Simultaneously, among the binary graphite intercalation compounds, YbC_6 and CaC_6 possess the highest critical temperatures, that reach respectively 6.5 and 11.5 K.

The whole of the results described in this chapter shows clearly that the intercalation and/or doping of carbon based materials is an excellent method to prepare novel superconductors. We have to bring this work to a successful conclusion.

REFERENCES

[1]   Hérold, C.; Marêché, J.-F.; Lagrange, P. *Carbon* 1996, 34, 517-521.
[2]   Legendre, A. Le matériau carbone ; *Eyrolles* : Paris, Fr, 1992.
[3]   Delhaes, P. in *Le carbone dans tous ses états*, Bernier, P.; Lefrant, S. Eds,
      Gordon and Breach Science Publishers : Amsterdam (NL), 1997, pp 41-82.
[4]   Mauguin, C. *Bull. Soc. Franç. Min.* 1926, 49, 32-61.
[5]   Enoki, T.; Suzuki, M.; Endo, M. Graphite Intercalation Compounds and
      Applications ; Oxford University Press : New York, 2003.
[6]   Hérold, A. *Chemical physics of intercalation*; series B: physics vol.172;
      NATO ASI series; Plenum Press New York and London, 1987, pp1-43.
[7]   Lagrange, P. *Chemical physics of intercalation II*; series B: physics vol.305;
      NATO ASI series; Plenum Press New York and London, 1993, pp 303-310.
[8]   Hérold, A. *Intercalation compounds of graphite*; Elsevier Sequoia:
      Lausanne, NL, 1977; vol. 31, pp 1-16.
[9]   Hérold, A. *Intercalated Materials*, Reidel Publishing compagny :
      Dordrecht, NL, 1979, pp 323-421.
[10] Hérold, A. *Bull. Soc. Chim. Fr.* 1955; 999-1012.
[11] Guérard, D. ; Hérold, A. *Carbon* 1975, 13, 337-345.
[12] Janot, R.; Guérard, D. *Prog. Mater. Sci.* 2005, 50, 1-92.
[13] Hérold, A.; Billaud, D.; Guérard, D.; Lagrange, P. *Intercalation compounds
      of graphite*; Elsevier Sequoia: Lausanne, NL, 1977; vol. 31, pp 25-28.
[14] Pruvost, S.; Hérold, C.; Hérold, A.; Lagrange, P. Eur. *J. Inorg. Chem.* 2004,
      1661-1667.
[15] Billaud, D.; Hérold, A. *Bull. Soc. Chim. Fr.* 1972, 103-107.
[16] Billaud, D.; Balesdent, D.; Hérold, A. *Bull. Soc. Chim. Fr.* 1974, 2402-
      2406.
[17] Billaud, D. ; Hérold, A. *Bull. Soc. Chim. Fr.* 1974, 2407-2410.
[18] Lagrange, P. *J. Mater. Res.* 1987, 2, 839-845.

[19] Hérold, C. ; Lagrange, P. C. R. *Chimie* 2003, 6, 457-465.

[20] Lagrange, P. ; Hérold, A. *Carbon* 1978, 16, 235-240.

[21] Hérold, A. *Mater. Sci. Eng.* 1977, 31, 1-16.

[22] Lagrange, P.; Guérard, D.; Hérold, A. *Ann. Chim. Fr.* 1978, 3, 143-159.

[23] Lagrange, P.; Guérard, D.; El Makrini, M.; Hérold A. C. R. *Acad. Sci. Paris série C* 1978, 287, 179-182.

[24] Guérard, D.; Lagrange, P.; El Makrini, M.; Hérold, A. *Carbon* 1978, 16, 285-290.

[25] Métrot, A.; Guérard, D.; Billaud, D.; Hérold, A. *Synth. Metals* 1979/80, 1, 363-369.

[26] Guérard, D.; Chaabouni, M.; Lagrange, P.; El Makrini, M.; Hérold, A. *Carbon*, 1980, 18, 257-264.

[27] Pruvost, S.; Hérold, C.; Hérold, A.; Lagrange, P. *Carbon* 2003, 41, 1281-1289.

[28] Pruvost, S.; Hérold, C.; Hérold, A.; Lagrange, P. *Carbon* 2004, 42, 1825-1831.

[29] Emery, N.; Hérold, C.; Lagrange, P. *J. Sol. St. Chem.* 2005, 9, 2947-2952.

[30] El Makrini, M.; Guérard, D.; Lagrange, P.; Hérold, A. *Carbon*, 1980, 18, 203-209.

[31] Guérard, D.; Nalimova, V. A. *Mol. Cryst. Liq. Cryst.* 1994, 244, 263-268.

[32] Nalimova, V. A.; Guérard, D.; Lelaurain, M.; Fateev, O. V. *Carbon* 1995, 33, 177-181.

[33] Avdeev, V. V.; Nalimova, V. A.; Semenenko, K. N. *Synth. Metals* 1990, 38, 363-369.

[34] Avdeev, V. V.; Nalimova, V. A.; Semenenko, K. N. *High Pressure Res.* 1990,6, 11-25.

[35] Nalimova, V. A.; Chepurko, S. N.; Avdeev, V. V.; Semenenko, K. N. *Synth. Metals*, 1991, 40, 267-273.

[36] Guérard, D.; Janot, R. *J. Phys. Chem. Sol.* 2004, 65, 147-152.

[37] Nalimova, V. A. *Mol. Cryst. Liq. Cryst.* 1998, 310, 5-17.

[38] Dresselhaus, M. S.; Dresselhaus, G. *Adv. Phys.* 1981, 30, 139-326.

[39] Kelly, B. T.; Physics of Graphite ; *Applied Science* : London UK, 1981.

[40] Blinowski, J.; Rigaux, C. *J. Phys.* (Paris) 1980, 41, 667-676.

[41] Blinowski, J.; Rigaux, C. *J. Phys.* (Paris) 1984, 45, 545-555.

[42] Holzwarth, N. A. W.; Girifalco, A. *Phys. Rev. B* 1978, 18, 5190-5205.

[43] Holzwarth, N. A. W.; Laurie, S. G.; Rabii S. *Phys. Rev. B* 1983, 28, 1013-1025.

[44] Kasowski, R. V. *Phys. Rev. B* 1982, 25, 4189-4195.

[45] Ohno, T. *J. Phys. Soc. Jpn Suppl. A* 1980, 49, 899-902.

[46] Posternak, M.; Balderschi, A.; Freeman, A. J.; Wimmer, E. *Phys. Rev. Lett.* 1983, 50, 761-764.

[47] Samuelson, L.; Batra, I. P. *J. Phys. C* 1980, 13, 5105-5124.

[48] Frétigny, C.; Saito, R.; Kamimura, H. *J. Phys. Soc. Jpn* 1989, 58, 2098-2108.

[49] Koma, A.; Miki, K.; Suematsu, H.; Ohno, T.; Kamimura, H. *Phys. Rev. B* 1986, 34, 2434-2438.

[50] Holzwarth, N. A. W.; Louie, S. E.; Rabii, S. *Phys. Rev. B* 1984, 30, 2219-2222.

[51] Posternak, M.; Balderschi, A.; Freeman, A. J.; Wimmer, E. *Phys. Rev. Lett.* 1983, 52, 863-866.

[52] Saito, M.; Oshiyama, A. *J. Phys. Soc. Jpn* 1986, 55, 4341-4348.

[53] Tatar, R. C. Ph. D. Thesis, University of Pennsylvania, 1985.

[54] Mizuno, S.; Hiramoto, H.; Nakao, K. *J. Phys. Soc. Jpn* 1987, 56, 4466-4476.

[55] Gunasekara, N.; Takahashi, T.; Maeda, F.; Sagawa, T.; Suematsu, H. Z. *Phys. B* 1988, 70, 349-355.

[56] Holzwarth, N. A. W.; DiVicenzo, D. P.; Tatar R. C.; Rabii, S. *Int. J. Quantum Chem.* 1983, 23, 1223-1230.

[57] Preil, M. E.; Fischer, J. E.; DiCenzo, S. B.; Werthheim, G. K. *Phys. Rev. B* 1984, 30, 3536-3538.

[58] Woo, K. C.; Flanders, P. J.; Fischer, J. E. *Bull. Am. Phys. Soc.* 1982, 27, 272-278.

[59] Prudnikova, G. V.; Gjatkin, A.; Ermakov, A. V.; Shinkin, A. M.; Adamchuk, V. K. J. Electron Spectrosc. *Relat. Phenom.* 1994, 68, 427-430.

[60] Shinkin, A. M.; Molodtsov, S. L.; Laubschat, C.; Kaindl, G.; Prudnikova, G. V.; Adamchuk, V. K. *Phys. Rev. B* 1995, 51, 13586-13591.

[61] Shinkin, A. M.; Prudnikova, G. V.; Adamchuk, V. K.; Molodtsov, S. L.; Laubschat, C.; Kaindl, G. *Surf. Sci.* 1995, 331-333, 517-521.

[62] Molodtsov, S. L.; Gantz, T.; Laubschat, C.; Viatkine, A. G.; Avila, J.; Cassdo, C.; Asensio, M. C. Z. *Phys. B* 1996, 100, 381-385.

[63] Molodtsov, S. L.; Laubschat, C.; Richter, M.; Gantz, T.; Shinkin, A. M. *Phy. Rev. B* 1996, 53, 16621-16630.

[64] Louie, S. G.; Cohen, M. L. *Phys. Rev. B* 1974, 10, 3237-3245.

[65] Parker, L. J.; Atou, T.; Badding, J. V. *Science* 1996, 273, 95-97.

[66] Krueger, C.; Tsay, Y.-H. *Angew. Chem.* 1973, 85, 1051-1052.

[67] Nalimova, V. A.; Semenenko, K. N.; Avdeev, V. V. *Synth. Metals* 1992, 46, 79-84.

[68] Delhaes, P. *Mater. Sci. Eng.* 1977, 31, 225-234.

[69] Suematsu, H.; Higuchi, K.; Tanuma, S. *J. Phys. Soc. Jpn* 1980, 48, 1541-1549.

[70] Potter, M. E.; Johnson, W. D.; Fischer, J. E. *Solid State Commun.* 1981, 37, 713-718.

[71] McRae, E.; Marêché, J.-F.; Pernot, P.; Vangelisti, R. *Phys. Rev. B* 1989, 39, 9922-9928.

[72] Issi, J.-P. in Graphite Intercalation Compounds II ; Zabel H., Solin S. A., Eds ; Springer-Verlag : Berlin, 1992, 195-245.

[73] Tsuzuku, T. *Carbon* 1979, 17, 293-299.

[74] Basu, S.; Zeller, C.; Flanders, P.; Fuerst, C. D.; Johnson, W. D.; Fischer, J. E. *Mater. Sci. Eng.* 1979, 38, 275-283.

[75] Murray, J. J.; Ubbelohde, A. R. *Proc. Roy. Soc. A* 1969, 312, 371-379.

[76] Sugihara, K. *Phys. Rev. B* 1988, 37, 4752-4759.

[77] Sugihara, K. *Phys. Rev. B* 1988, 37, 7063-7069.

[78] Sugihara, K.; Chen, N. C.; Dresselhaus, M. S.; Dresselhaus, G. *Phys. Rev. B* 1989, 39, 4577-4587.

[79] Shimamura, S. *Synth. Metals* 1985, 12, 365-370.

[80] Hannay, N.B.; Geballe, T.H.; Matthias, B.T.; Andres, K.; Schmidt, P.; Mac Nair, D. *Phys. Rev. Lett.* 1965, 14, 225-226.

[81] Koike, Y.; Suematsu, H.; Higuchi, K.; Tanuma, S. *Solid State Comm.* 1978, 27, 623-627.

[82] Koike, Y.; Tanuma, S. *J. Phys. Soc. Jpn.* 1981, 50, 1964-1969.

[83] Koike, Y.; Tanuma, S.; Suematsu, H.; Higuchi, K. *J. Phys. Chem. Solids* 1980, 41, 1111-1118.

[84] Kobayashi, M.; Tsujikawa, I.J. *Physica B* 105, 1981, 439-443.

[85] Kobayashi, M.; Tsujikawa, I.J. *J. Phys. Soc. Jpn.* 1981, 50, 3245-3253.

[86] Kobayashi, M.; Enoki, T.; Inokuchi, H.; Sano, M.; Sumiyama, A.; Oda, Y.; Nagano, H. *Synth Metals* 1985, 12, 341-346.

[87] Weller, T.E.; Ellerby, M.; Saxena, S. S.; Smith, R.P.; Skipper, N.T. *Nat. Phys.* 2005, 1, 39-41.

[88] Emery, N.; Hérold, C.; d'Astuto, M.; Garcia, V.; Bellin, C.; Marêché, J.F.; Lagrange, P.; Loupias, G. *Phys. Rev. Lett.* 2005, 95, 087003.

[89] Kaneiwa, S.; Kobayashi, M.; Tsujikawa, I. *J. Phys. Soc. Jpn* 1982, 51, 2375-2376.

[90] DeLong, L.E.; Yeh, V.; Tondiglia, V.; Eklund, P.C.; Lambert S.E.; Maple M.B. *Phys. Rev. B* 1982, 26, 6315-6318.

[91] DeLong, L.E.; Eklund, P.C. *Synth. Metals* 1983, 5, 291-300.

[92] Belash, I.T.; Bronnikov, A.D.; Zharikov, O.V.; Pal'nichenko, A.V. *Solid state Comm.* 1987, 63,153-155.

[93]   Belash, I.T.; Zharikov, O.V.; Pal'nichenko, A.V. *Synth. Metals*, 1989, 34, 455-460.

[94]   Belash, I.T.; Bronnikov, A.D.; Zharikov, O.V.; Pal'nichenko, A.V. *Solid State comm.* 1989, 69, 921-923.

[95]   Belash, I.T.; Bronnikov, A.D.; Zharikov, O.V.; Pal'nichenko, A.V. *Synth. Metals* 1990, 36, 283-302.

[96]   Belash, I.T.; Bronnikov, A.D.; Zharikov, O.V.; Pal'nichenko, A.V. *Solid state Comm.* 1987, 64, 1445-1447.

[97]   Avdeev, V.V.; Zharikov, O.V.; Nalimova, V.A.; Pal'nichenko, A.V.; Semenenko, K.N.; *Zh. Eksp. Teor. Fiz.* 1986, 43, 376-378.

[98]   Nalimova, V.A.; Avdeev, V.V.; Semenenko, K.N. *Mater. Sci. Forum* 1992, 91-93, 11-16.

[99]   Calandra, M.; Mauri, F. *Phys Rev Lett.* 2005, 95, 237002.

[100] Calandra, M.; Mauri, F. *Phys. Stat. Sol.* (b), 2006, 243, 3458-3463.

[101] Mazin, I I. *Phys Rev Lett.* 2005, 95, 227001.

[102] Mazin, I.I.; Molodtsov, S.L. *Phys. Rev. B*, 2005, 72, 172504.

[103] Csányi, G.; Littlewood, P.B.; Nevidomskyy, A.H.; Pickard, C.J.; Simons, B.D. *Nat. Phys.* 2005, 1, 42-45.

[104] Ellerby, M.; Weller, T.E.; Saxena, S.S.; Smith, R.P.; Skipper, N.T. *Physica B* 2006, 378-380, 636-639.

[105] Xie, R.; Rosenmann, D.; Rydh, A.; Claus, H.; Karapetrov, G.; Kwok, W.K.; Welp, U. *Physica C* 2006, 439, 43-46.

[106] Jobiliong, E.; Zhou, H. D.; Janik, J.A.; Jo, Y.-J.; Balicas, L.; Brooks, J.S.; Wiebe, C.R. *Phys. Rev. B* 2007, 76, 052511.

[107] Lamura, G.; Aurino, M.; Cifariello, G.; Di Gennaro, E.; Andreone, A.; Emery, N.; Hérold, C.; Marêché, J.-F.; Lagrange, P. *Phys. Rev. Let.* 2006, 96, 107008.

[108] Lamura, G.; Aurino, M.; Cifariello, G.; Di Gennaro, E.; Andreone, A.; Emery, N.; Hérold, C.; Marêché, J.-F.; Lagrange, P. *Physica C* 2007, 714-715.

[109] Gauzzi, A.; Cochec, J.L.; Lamura, G.; Jönsson, B.J.; Gasparov, V.A.; Ladan, F.R.; Plaçais, B.; Probst, P.A.; Pavuna, D.; Bok, *J. Rev. Sci. Instrum.* 2000, 71, 2147.

[110] Marsiglio, F.; Carbotte, J.P.; Blezius, *J. Phys. Rev. B* 1990, 41, 6457.

[111] Cifariello, G.; Di Gennaro, E.; Lamura, G.; Andreone, A.; Emery, N.; Hérold, C.; Marêché, J.-F.; Lagrange, P. *Physica C* 2007, 716-717.

[112] Andreone, A.; Cifariello, G.; Di Gennaro, E.; Lamura, G.; Emery, N.; Hérold, C.; Marêché, J.-F.; Lagrange P., *Appl. Phys. Lett.* 2007, 91, 072512.

[113] Kim, J.S.; Kremer, R.K.; Boeri, L.; Razavi, F.S. *Phys. Rev. Lett.* 2006, 96, 217002.

[114] Mazin, I.I.; Boeri, L.; Dolgov, O.V.; Golubov, A.A.; Bachelet, G.B.; Giantomassi, M.; Andersen, O.K. *Physica C* 2007, 460-462 116-120.

[115] Sanna, A.; Profeta, G.; Floris, A.; Marini, A.; Gross, E.K.U.; Massidda, S. *Phys. Rev. B*, 2007, 75, 020511R.

[116] Bergeal, N.; Dubost, V.; Noat, Y.; Sacks, W.; Roditchev, D.; Emery, N.; Hérold, C.; Marêché, J.-F.; Lagrange, P.; Loupias, G. *Phys. Rev. Lett.* 2006, 97, 077003.

[117] Kurter, C. ; Oyuzer, L. ; Mazur, D. ; Zasadzinski, J.F. ; Rosenmann, D. ; Claus, H. ; Hinks, D.G. ; Gray, D.E. *Phys. Rev. B* 2007, 76, 220502R.

[118] Gonelli, R.S. ; Delaude, D. ; Tortello, M. ; Ummarino, G.A. ; Stepanov, V.A. ; Kim, J.S. ; Kremer, R.S. ; Sanna, A. ; Profeta, G. ; Massidda, A. *Cond-Mat.* 2007, 0708.0921.

[119] Nagel, U. ; Hüvonen, D. ; Joon, E. ; Kim, J.S. ; Kremer, R.K. ; Rõõm, T. *Phys Rev B* 2008, 78, 041404R.

[120] Hlinka, J. ; Gregora, I. ; Pokorny, J. ; Hérold, C. ; Emery, N. ; Marêché, J.F. ; Lagrange, P. *Phys Rev B* 2007, 76, 144512.

[121] Hinks, D.G. ; Rosenmann, D. ; Claus, H. ; Bailey, M.S. ; Jorgensen, J.D. *Phys. Rev. B* 2007, 75, 014509.

[122] Deng, S.; Simon, A.; Köhler, J. *Angew. Chem. Int.* 2008, 47, 6703.

[123] Hinks, D.G.; Rosenmann, D.; Claus, H.; Bailey, M.S.; Jorgensen, J.D. *Phys. Rev. B* 2007, 75, 014509.

[124] Kim, J.S.; Boeri, L.; O'Brien, J.R.; Razavi, F.S.; Kremer, R.K. *Phys. Rev. Lett.*, 2007, 99, 027001.

[125] Nakamae, S.; Gauzzi, A.; Ladieu, F.; L'Hôte, D.; Emery, N.; Hérold, C.; Marêché, J.F.; Lagrange, P.; Loupias, G., *Solid State Comm* 2008, 145, 493-496.

[126] Calandra, M.; Mauri, F., *Phys Rev B* 2006, 74, 094507.

[127] Sutherland, M.; Doiron-Leyraud, N.; Taillefer, L.; Weller, T.; Ellerby, M.; Saxena, S.S. *Phys. Rev. Lett.* 2007, 98, 067003.

[128] Smith, R.P.; Weller, T.E.; Kusmartseva, A.F.; N.T. Skipper, N.T.; Ellerby, M; Saxena, S.S. *Physica B* 2006, 378-380, 992-993.

[129] Smith, R.P.; Kusmartseva, A.; Ko, Y.T.C.; Saxena, S.S.; Akrap, A.; Forro, L.; Laad, M.; Weller, T.E.; Ellerby, M.; Skipper, N.T. *Phys. Rev. B* 2006, 74, 024505.

[130] Clarke, R.; Uher, C. Adv. Phys. 1984, 33, 469-566.

[131] Kim, J.S.; Boeri, L.; Kremer, R.K.; Razavi, F.S. *Phys. Rev. B* 2006, 74, 214513.

[132] Gauzzi, A.; Takashima, S.; Takeshita, N.; Terakura, C.; Takagi, H.; Emery, N.; Hérold, C. ; Lagrange, P. ; Loupias, G. *Phys. Rev. Lett.* 2007, 98, 067002.

[133] Zhang, L.; Xie, Y.; Cui, T.; Li, Y.; He, Z.; Ma, Y.; Zou, G. *Phys. Rev.*, B 2006, 74, 184519.

[134] Gauzzi, A. ; Bendiab, N. ; d'Astuto, M. ; Canny, B. ; Calandra, M. ; Mauri, F. ; Loupias, G. ; Emery, N. ; Hérold, C. ; Lagrange, P. ; Hanfland, M. ; Mezouar, M. Phys. Rev. B 78 (2008) 064506

[135] Csanyi, G.; Pickard, C.J.; Simons, B.D.; Needs, R.J. *Phys. Rev. B* 2007, 75, 085432.

[136] Nagamatsu, J.; Nakagawa, N.; Muranaka, T.; Zenitani, Y.; Akimitsu, J. *Nature*, 2001, 410, 63-64.

[137] Canfield, P.C.; Bud'ko, S.L.; Finnemore, D.K. *Physica C,* 2003, 385, 1-7.

[138] Iavarone, M.; Karapetrov, G.; Koshelev, A.E.; Kwok, W.K.; Crabtree, G.W.; Hinks, D.G.; Kang, W.N.; Choi, E-M.; Kim, H.J.; Kim, H.-J.; Lee, S.I. *Phys. Rev. Lett.*, 2002, 89, 187002.

[139] Wosylus, A. ; Prots, Y. ; Burkhardt, U. ; Schnelle, W. ; Schwarz, U. *Sci and Techn Adv Mat* 2007, 8, 383-388.

[140] El Makrini, M.; Lagrange, P.; Guérard, D.; Hérold, A. *Carbon*, 1980, 18, 211-216.

[141] Lagrange, P.; El Makrini, M.; Guérard, D.; Hérold, A. *Physica B* 1980, 99, 473-476.

[142] Lagrange, P.; Outti, B.; Assouik, J.; Clément J. *Mater. Sci. Forum* 1992, 91-93, 295-299.

[143] Outti, B.; Clément, J.; Hérold, C.; Lagrange, P. *Mol. Cryst. Liq. Cryst.* 1994, 244-245, 281-286.

[144] Emery, N.; Pruvost, S.; Hérold, C.; Lagrange, P. *J. Phys. Chem. Solids* 2006, 67, 1137-1140.

[145] Hérold, C.; Pruvost, S.; Hérold, A.; Lagrange, P. *Carbon* 2004, 42, 2113-2115.

[146] Lagrange, P.; Bendriss-Rerhrhaye, A.; Marêché, J.-F.; McRae, E. *Synth. Metals* 1985, 12, 201-206.

[147] Bendriss-Rerhrhaye, A.; Lagrange, P.; Rousseaux, F. *Synth. Metals* 1988, 23, 89-94.

[148] Essaddek, A.; Lagrange, P.; Rousseaux, F. *Synth. Metals* 1989, 34, 255-260.

[149] Essaddek, A.; Lelaurain, M.; Marêché, J.-F.; McRae, E.; Lagrange, P. *Synth. Metals* 1989, 34, 365-370.

[150] Assouik, J.; Lagrange, P. *Mater. Sci. Forum* 1992, 91-93, 313-317.

[151] Hérold, C.; Goutfer-Wurmser, F.; Lagrange, P. *Mol. Cryst. Liq. Cryst.* 1998, 310,57-62.

[152] Saehr, D.; Hérold, A. *Bull. Soc. Chim. Fr.* 1965, 1965, 3130-3136.

[153] Colin, M.; Hérold, A. *Bull. Soc. Chim. Fr.* 1971, 1971, 1982-1990.

[154] Guérard, D.; Lagrange, P.; Hérold, A. *Mater. Sci. Eng.* 1977, 31, 29-32.

[155] Hérold, A.; Lagrange, P. *Mater. Sci. Eng.* 1977, 31, 33-37.

[156] Guérard, D.; Takoudjou, C.; Rousseaux, F. *Synth. Metals* 1983, 7, 43-48.

[157] Guérard, D.; Elalem, N. E.; Takoudjou, C.; Rousseaux, F. *Synth. Metals* 1985, 12, 195-200.

[158] Guérard, D., Elalem, N. E.; El Hadigui, S.; Elansari, L.; Lagrange, P.; Rousseaux, F.; Estrade-Szwarckopf, H.; Conard, J.; Lauginie, P. *J. Less-Common Metals* 1987, 131, 173-180.

[159] Guérard, D.; Elansari, L.; Elalem, N. E.; Marêché, J.-F.; McRae, E. *Synth. Metals* 1990, 34, 27-32.

[160] El Gadi, M.; Hérold, A.; Hérold, C.; Marêché, J.-F.; Lagrange, P., *J. Sol. St. Chem.*, 1997, 131, 282-289.

[161] El Gadi, M.; Hérold, C.; Lagrange, P. *Carbon* 1994, 32, 749-752.

[162] Hérold, C.; El Gadi, M.; Marêché, J.-F.; Lagrange, P. *Mol. Cryst. Liq. Cryst.* 1994, 244, 41-46.

[163] Lagrange, P.; Hérold, C.; Nalimova, V. A.; Sklovsky, D. E.; Guérard, D. *J. Phys. Chem. Solids* 1996, 57, 707-713.

[164] El Gadi, M.; Hérold, C.; Lagrange, P. *C. R. Acad. Sci. Serie II* 1993, 316, 763-769.

[165] Yamashita, T.; Mordkovich, V. Z.; Murakami, Y.; Suematsu, H.; Enoki, T. *J. Phys. Chem. Solids*, 1996, 57, 765-769.

[166] Mordkovich, V. Z.; Baxendale, M.; Ohki, Y.; Yoshimura, S.; Yamashita, T.; Enoki, T. *J. Phys. Chem. Solids* 1996, 57, 821-825.

[167] Goutfer-Wurmser, F.; Hérold, C.; Marêché, J.-F.; Lagrange, P. *Mol.Cryst. Liq. Cryst.* 1998, 310, 51-56.

[168] Goutfer-Wurmser, F.; Hérold, C.; Lagrange, P. *Ann. Chim. Sci. Mat.* 2000, 25, 101-118.

[169] Hérold, A.; Lelaurain, M.; Marêché, J.F.; McRae, E. *C. R. Acad. Sci. Paris Ser. II*, 1995, 321, 61-67.

[170] Hérold, A.; Marêché, J.-F.; Lelaurain, M. *Carbon* 2000, 38, 1955-1963.

[171] Yang, M. H.; Eklund, P. C. *Phys. Rev. B* 1988, 38,3505-3516.

[172] Yang, M. H.; Charron, P. A.; Heinz, R. E.; Eklund, P. C. *Phys. Rev. B* 1988, 37, 1711-1718.

[173] Enoki, T.; Miyajima, S.; Sano, M.; Inokuchi, H. *J. Mater. Res.* 1990, 5, 435-466.

[174] Miyajima, S.; Kabasawa, M.; Chiba, T.; Enoki, T.; Maruyama, Y.; Inokuchi, H. *Phys. Rev. Lett.* 1990, 64, 319-322.

[175] Mizuno, S.; Nakao, K. *Phys. Rev. B* 1990, 41, 4938-4947.

[176] Enoki T. In Supercarbon, Yoshimura S., Chang R. P. H. Eds. ; Springer : Berlin, 1998, p. 137.

[177] Higai, S.; Mizuno, S.; Nakao, K. *J. Phys. Chem. Solids* 1996, 57, 689-694.

[178] Rodriguez-Fortea, A.; Rovira, C.; Ordejon, P.; Hérold, C.; Lagrange, P.; Canadell, E. *Inorg. Chem.* 2006, 45, 9387-9393.

[179] Marêché, J.-F.; McRae, E.; Bendriss-Rerhrhaye, A.; Lagrange, P. *J. Phys. Chem. Solids* 1986,47, 477-483.

[180] Alexander, M.G.; Goshron, D.P.; Guérard, D.; El Makrini, M.; Lagrange, P.; Onn, D.G. *Synth. Metals* 1980, 2, 203-211.

[181] Pendrys, L.A.; Wachnik, R.; Vogel, F.L.; Lagrange, P.; Furdin, G.; El Makrini, M.; Hérold, A. *Solid State Comm.* 1981, 38, 677-678.

[182] Koike, Y.; Tanuma, S.I. *J. Phys. Soc. Japan* 1981, 50, 1964-1969.

[183] Alexander, M.G.; Goshorn, D.P.; Guérard, D.; Lagrange, P.; El Makrini M.; Onn, D.G. *Solid State Comm.* 1981, 38, 103-107.

[184] Iye Y.; Tanuma, S.I.; *Phys. Rev. B* 1982, 25, 4583-4592.

[185] Chaiken, A.; Dresselhaus, M.S.; Orlando, T.P.; Dresselhaus, G.; Tedrow, P.M.; Neumann, D.A.; Kamitakahara W.A. *Phys. Rev. B* 1990, 41, 71-81.

[186] Tsukamoto, J.; Fukuda, D.; Takahashi, A.; Murata, K.; *Appl. Phys. Lett.* 1989, 55, 2035-2036.

[187] Timp, G.; Elman, B.S.; Dresselhaus, M.S.; Tredow, P. *Proc. Mat. Res. Soc. Symp.* 1983, 20, 201-206.

[188] Roth, G.; Chaiken, A.; Enoki, T.; Yeh, N.C.; Dresselhaus, G.; Tedrow, M.P. *Phys. Rev. B* 1985, 32, 533-536.

[189] Wachnik, R.A.; Pendrys, L.A.; Vogel, F.L.; Lagrange P. *Solid State Comm.* 1982,43, 5-8.

[190] Emery, N.; Hérold, C.; Marêché, J.-F.; Bellouard, C.; Loupias, G.; Lagrange, P. *J. Solid State Chem.* 2006, 179, 1289-1292.

[191] Iye, Y.; Tanuma, S.I. *Synth. Metals* 1983, 5, 257-276.

[192] Pendrys, L.A.; Wachnik, R.A.; Vogel, F.L.; Lagrange, P. *Synth. Metals*, 1983, 5, 277-290.

[193] Mc Rae, E.; Marêché, J.F.; Bendriss-Rerhrhaye, A.; Lagrange, P.; Lelaurain, M. *Annales de Physique*, Colloque 1986, 11, 13-22.

[194] Stang, I.; Lüders, K.; V.; Güntherodt, H.J. *Synt. Metals* 1988, 23, 371-375.

[195] Lüders, K. *Chemical physics of intercalation II*; series B: physics vol.305; NATO ASI series; Plenum Press New York and London, 1993, pp 31-61.

[196] Ekimov, E.A.; Sidorov, V.A.; Bauer, E.D.; Mel'nik, N.N.; Curro, N.J.; Thompson, J.D.; Stishov, S.M. *Nature*, 2004, 428, 542-545.

[197] Morpurgo, A. F.; Kong, J.; Marcus, C. M.; Dai, H. *Science* , 1999, 286, 263-265.

[198] Kasumov, A. Yu.; Deblock, R.; Kociak, M.; Reulet, B.; Bouchiat, H.; Khodos, I. I.; Gorbatov, Yu. B.; Volkov, V. T.; Journet, C.; Burghard, M. *Science*, 1999, 284, 1508-1511.

[199] Kociak, M.; Kasumov, A. Yu.; Guéron, S.; Reulet, B.; Khodos, I.I.; Gorbatov, Yu. B.; Volkov, B. T.; Vaccarini, L.; Bouchiat, H. *Phys. Rev. Lett.* 2001, 2416.

[200] Tang Z. K.; Zhang L.; Wang N.; Zhang X. X.; Wen G. H.; Li G. D.; Wang J. N.; Chan C. T.; Sheng P. *Science*, 2001, 292, 2462-65.

[201] Takesue, I.; Haruyama, J.; Kobayashi, N.; Chiashi, S. Maruyama, S.; Sugai T.; Shinohara, H. *Phys. Rev. Lett.* 2006, 057001.

[202] Hebard, A.F.; Rosseinsky, M.J.; Haddon, R.C.; Murphy, D.W.; Glarum, S.H.; Palstra, T.T.M.; Ramirez, A.P.; Kortan, A.R. *Nature* 1991, 350, 600-601.

[203] Holczer, K.; Klein, O.; Grüner, G.; Huang, S-M.; Kaner, R.B.; Fu, J.K.; Whetten, R.L.; Dietrich, F. *Science* 252, 1991, 1154-1157.

[204] Fleming, R.M.; Ramirez, A.P.; Rosseinsky, M.J.; Murphy, D.W.; Haddon, R.C.; Zahurak, S.M.; Makhija, A.V. *Nature* 1991, 352, 787-788.

[205] Tanigaki, K.; Ebbesen, T.W.; Saito, S.; Mizuki, J.; Tsai, J.S.; Kubo, Y.; Kuroshima, S. *Nature* 1991, 352, 222-223.

[206] Hérold, C.; Marêché, J.F.; Lagrange, P. *C. R. Acad. Sci. Paris*, 1995, 321, Série IIb, 103-110.

[207] Palstra, T.T.M.; Zhou, O.; Iwasa, Y.; Sulewski, P.E.; Fleming, R.M.; Zegarski, B.R. *Solid State Comm.* 1995, 93, 327-330.

[208] Tanigaki, K.; Ebbesen, T.W.; Tasi, J.S.; Hirosawa, I.; Mizuki, J. *Europhys. Lett.*, 1993, 23, 57-62.

[209] Margadonna, S.; Prassides, K. *J. Solid State Chem.* 2002, 168, 639-652.

[210] Zhou, O.; Fleming, R.M.; Murphy, D.W.; Rosseinsky, M.J.; Ramirez, A.P.; van Dover, R.B.; Haddon, R.C. *Nature*, 1993, 362, 433-435.

[211] Kortan, A.R.; Kopylov, N.; Glarum, S.; Gyorgy, E.M.; Ramirez, A.P.; Fleming, R.M.; Thiel, F.A.; Haddon, R.C. *Nature*, 1992, 355, 529-532.

[212] Kortan, A.R.; Kopylov, N.; Glarum, S.; Gyorgy, E.M.; Ramirez, A.P.; Fleming, R.M.; Zhou, O.; Thiel, F.A.; Trevor, P.L.; Haddon, R.C. *Nature*, 1992, 360, 566-568.

[213] Ozdas, E.; Kortan, A.R.; Kopylov, N.; Ramirez, A.P.; Siegrist, T.; Rabe, K.M.; Bair, H.E.; Schuppler, S.; Citrin, P.H. *Nature*, 1995, 375, 126-129.

[214] Chen, X.H.; Roth, G. *Phys. Rev. B,* 1995, 52, 15534.

[215] Akada, M.; Hirai, T.; Takeuchi, J.; Yamamoto, T.; Kumashiro, R.; Tanigaki, K. *Phys. Rev. B*, 2006, 73, 094509.

INDEX

D

E